高职高专行业英语系列教材　ESP

计算机英语

(第3版)

主编　钟文龙

重庆大学出版社

内容提要

《计算机英语》系高职高专行业英语系列教材之一，是一本融合了计算机知识和实用英语双领域的教材。本书共有十个单元和四个附录，分别介绍了计算机的发展历史、计算机的组成、计算机的操作系统、数据库系统、编程语言、计算机网络技术和互联网技术、信息安全、多媒体技术和电脑游戏、电子商务技术、计算机未来的发展方向和最新的技术概念等。本书既可作为高等职业院校计算机类专业学习行业英语的教材，也可以作为从事其他专业的读者了解计算机原理、科学知识、认识计算机技术的应用和发展的辅助参考手册。

图书在版编目(CIP)数据

计算机英语：英文／钟文龙主编．--3版．—重庆：重庆大学出版社，2020.8
高职高专行业英语系列教材
ISBN 978-7-5689-2194-7

Ⅰ.①计… Ⅱ.①钟… Ⅲ.①电子计算机—英语—高等职业教育—教材 Ⅳ.①TP3

中国版本图书馆CIP数据核字（2020）第094089号

计算机英语（第3版）

主编 钟文龙

责任编辑：陈 亮　　版式设计：安 娜
责任校对：陈 力　　责任印制：赵 晟

＊

重庆大学出版社出版发行
出版人：饶帮华
社址：重庆市沙坪坝区大学城西路21号
邮编：401331
电话：（023）88617190　88617185（中小学）
传真：（023）88617186　88617166
网址：http://www.cqup.com.cn
邮箱：fxk@cqup.com.cn（营销中心）
全国新华书店经销
重庆升光电力印务有限公司印刷

＊

开本：787mm×1092mm　1/16　印张：12.25　字数：384千
2020年8月第3版　2020年8月第5次印刷
ISBN 978-7-5689-2194-7　定价：46.00元

本书如有印刷、装订等质量问题，本社负责调换
版权所有，请勿擅自翻印和用本书
制作各类出版物及配套用书，违者必究

Unit 1

History of Computer

After reading this unit and completing the exercises, you will be able to

☆ understand the earliest calculating devices, such as abacus, Napier's Bones and so on;

☆ be familiar with the history of computer development;

☆ identify all the general elements of a computer structure;

☆ apply your knowledge when using your computers.

Part I Lead-in

1. *Look at the pictures and have a discussion about them. Then produce the possible words and expressions that are related to the theme.*

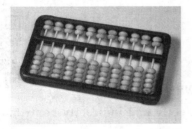
Picture 1

Picture 2

Picture 3

Picture 4

Write down the relevant words and expressions in the space provided below.

_____

_____

2. ***Listen to the passage and then fill in the blanks with the exact words or phrases you have just heard.***

  Using a computer is now part of _____ life for most Chinese, whether we like it or not! In fact, in today's China, you will inadvertently use a computer when purchasing a train or _____ ticket, paying for taxes, making a _____, watching an LCD television and such.

  Computers have an _____ upon our lives almost everywhere we do (or don't) go. Perhaps amazingly, computers are a relevantly recent _____.

Part II Read and Understand

1. ***Discuss the questions with your partners.***

☆ Where did the first mechanical calculator firstly appeared?

☆ What is the mechanical computer?

☆ What is the first electronic computer and when it was created?

☆ When did the first personal computer appear?

2. ***Time to read it.***

↪ The First Computer ↩

  Many discoveries and inventions have directly and indirectly contributed to the development of the personal computer. Examining a few important developmental landmarks can help bring entire picture into focus.

  The first computers of any kind were simple calculators. Even these evolved from mechanical devices to electronic digital devices.

前　言

随着计算机应用的日益普及和计算机技术的高速发展，人类已经进入了"信息时代"，现在正迈向"人工智能时代""5G时代"。在这个时代里，英语众望所归地成了通用语言，日新月异的信息技术不断产生大量的新名词，以及创新的表达方式，其知识更新突出地体现在计算机领域的技术书刊、杂志、资料等出版发行物上。掌握最新的信息资料和技术是对计算机类专业人才的基本要求。因此我们结合实际情况更新过时的内容，及时推出《计算机英语》（第3版）。

在计算机这个高新技术行业中，绝大部分新技术文献是欧美的技术专家在国际知名的期刊或网站上用英文发表的。欲成为未来计算机行业的优秀人才，必须具有熟练阅读这些最新技术文献的能力，以便及时了解计算机发展的前沿技术和信息。

计算机专业英语是高职高专院校计算机类专业学生的重要工具课程，它是一门综合了计算机专业知识和英语应用能力的专业课程。本课程旨在使学生掌握实用的专业英语词汇和计算机的基本概念，为阅读计算机专业文献和书籍打下坚实的基础，同时为在以后工作中解决与计算机专业英语相关的问题提供必要的知识保证。

本书在延续前两版编写风格的基础上，对教材内容进行了更新，删去了过时内容，同时修正了错误的地方，增加了最新的主流计算机技术的相关内容。内容涵盖计算机硬件基础知识、因特网与电子商务、智能手机与平板电脑技术、计算机软件技术等专业知识。本书从培养高素质技能型人才的目标出发，着重于培养读者实际应用英语的能力。通晓计算机技术的读者应能熟练掌握计算机英语词汇、语法结构，能阅读、翻译计算机英文资料，顺利进行网上业务交际，并具有相应英语知识。本书也可以作为从事其他专业的读者了解计算机原理、科学知识，认识计算机技术的应用和发展的辅助参考手册。

一本难易适中的教材是在学习上取得良好效果的关键，教材取材过难会挫伤学生的自信心，太易则会索然寡味。本书的编写目的就是为目前的高职院校计算机类专业的学生或开始自学计算机专业技术的人员提供一本与他们基础相适应的专业英语教材。

随着近几年来国务院和教育部大力发展高职高专教育的不断推进，在这样的大背景下本书的编写指导思想进行了大胆的革新，强调"理论够用，重在培养实践动手技术能力"的原则。在文章取材上尽量缩减篇幅，词汇量控制在较低水平上，并注重提高选材的实用性。总的原则是兼顾介绍计算机专业英语常用的基本词汇、术语和最大限度地适应目前高职高专学生的基础现状。

本书还有以下几个方面的特点：计算机专业知识丰富，介绍了必要的语法知识及专业文章的翻译方法与技巧，注意与其他计算机专业课程内容的衔接与知识补充，阅读材料难度适当，强调理解及分析，每单元配有关键词、注释及大量习题。附录中列出了常用的计算机专业词汇、网络聊天词汇、常见的计算机屏幕英语等。

本书共分为十个单元和四个附录。第一单元介绍计算机的发展历史，主要介绍了第一台计算机的由来和个人计算机的发展历史，并在辅助读物中简单介绍了著名的摩尔定律。第二单元详细介绍了计算机的组成，包括 CPU、主板、硬盘、显示器、打印机等硬件设备。第三单元主要介绍计算机的操作系统，包括广泛应用的 Windows 系列操作系统，以及目前流行的 iOS 和 Android 两种智能手机操作系统。第四单元介绍了数据库系统，包括文件数据库、层次数据库、网络数据库、关系型数据库等多种数据库模型，以及一些主流的数据库管理系统。第五单元介绍编程语言，包括主流的 VB、C、C++、Java 等编程语言。第六单元介绍现在流行的计算机网络技术和互联网技术，包括 TCP/IP、HTTP、WWW 等网络协议，以及局域网、广域网、城域网等网络结构。第七单元介绍信息安全，包括加密、验证、数字签名、防火墙、入侵检测系统等安全技术和基本的信息安全概念。第八单元介绍多媒体技术和电脑游戏。第九单元介绍电子商务技术，包括 B2B、B2C、C2C 等应用模式。第十单元介绍计算机未来的发展方向和最新的技术概念，包括人工智能、虚拟实现技术等。附录 A 介绍计算机专业英语及翻译技巧；附录 B 介绍计算机应用过程中经常出现的屏幕提示信息；附录 C 为全书的基本词汇索引；附录 D 介绍计算机英语中的常见缩略术语。本书由重庆航天职业技术学院的钟文龙老师主编。钟老师负责全书的大纲编写、统稿、修改、校正和定稿工作，同时编写了第一到第十单元以及附录等内容。

计算机技术方兴未艾，行业英语教学活动正在日益广泛和不断深入。由于计算机技术发展突飞猛进，新知识新名词层出不穷，加上编者水平有限，时间仓促，书中难免存在不足之处，恳请广大读者批评指正。

编　者
2020 年 5 月

Mechanical Calculators

One of the earliest calculating devices on record was the abacus, which has been known and widely used for more than 2,000 years. The abacus was a simple wooden rack holding parallel rods on which beads were strung. When these beads were manipulated back and forth according to certain rules, several different types of arithmetic operations could be performed.

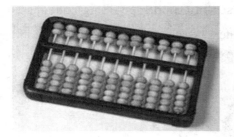 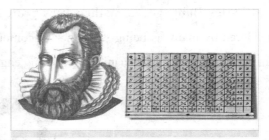

Math with standard Arabic numbers found its way to Europe in the eighth and ninth centuries. In the early 1600s a man named Charles Napier (the inventor of logarithms) developed a series of rods (later called Napier's Bones) that could be used to assist with numeric multiplication.

Blaise Pascal was normally credited with building the first digital calculating machine in 1642. It could perform the addition of numbers entered on dials and was intended to help his father, who was a tax collector. Then in 1671, Gottfried Wilhelm Von Leibniz invented a calculator that was finally built in 1694. His calculating machine could not only add, but by successive adding and shifting, could also multiply.

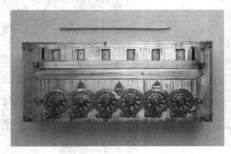

In 1820, Charles Xavier Thomas developed the first commercially successful mechanical calculator that could not only add but also subtract, multiply and divide. After that, a succession of ever improving mechanical calculators created by various other inventors followed.

The First Mechanical Computer

Charles Babbage, a mathematics professor in Cambridge, England, has been considered by many as being the father of computers because of his two great inventions—each a different type of mechanical computing engine.

The Difference Engine, as he called it, was conceived in 1812 and solved polynomial equations by the method of differences. By 1822, he had built a small working model of his Difference Engine for demonstration purposes. With financial help from the British government, Babbage started construction of full-scale model in 1823. It was intended to be steam-powered and fully automatic and would even print the resulting tables.

Babbage continued to work on it for 10 years, but by 1833 he had lost interest because he now had an idea for an even better machine, something he described as a general-purpose, fully program-controlled, automatic mechanical digital computer. Babbage called his new machine an Analytical Engine.

This Analytical Engine would have been the first true general-purpose computing device. It has been regarded as the first real predecessor to a modern computer because it had all the elements of what is considered a computer today. This included

☆ an input device (using an idea similar to the looms used in textile mills at the time, a form of punched cards supplied the input);

☆ a control unit (a barrel-shaped section with many slats and studs was used to control or program the processor);

☆ a processor (or calculator) (a computing engine containing hundreds of axles and thousands of gears about 10 feet tall);

☆ storage (a unit containing more axles and gears that could hold one thousand 50-digit numbers);

☆ an output device (plate designed to fit in a printing press, used to print the final results).

Electronic Computer

A physicist named John V. Atanasoff was credited with creating the first true digital electronic computer in 1942, while he worked at Iowa State College. His computer was the first to use modern digital switching techniques and vacuum tubes as the switches.

Military needs during World War II caused a great thrust forward in the evolution of computers. Systems were needed to calculate weapons trajectory and other military functions. In 1946, John P. Eckert, John W. Mauchly, and their associates in the Moore School of Electrical Engineering at the University of Pennsylvania built the first large scale electronic computer for the military. This machine became known as ENIAC, the Electrical Numerical Integrator and Calculator. It operated on 10-digit numbers and could multiply two such numbers at the rate of 300 products per second by finding the value of each product from a multiplication table stored in its memory. ENIAC was about 1,000 times faster than the previous generation of electromechanical relay computers.

ENIAC used about 18,000 vacuum tubes, occupied 1,800 square feet (167 square meters) of floor space, and consumed about 180,000 watts of electrical power. Punched cards served as the input and output; registers served as adders and also as quick-access read-write storage.

From ENIAC to the present, computer evolution has moved very rapidly. The first-generation computers were known for using vacuum tubes in their construction. The generation to follow would use the much smaller and more efficient transistor.

➵ Personal Computer ➵

A personal computer or PC is a small, relatively inexpensive computer designed for an individual user. In price, personal computers range anywhere from a few hundred dollars to over five thousand dollars. All are based on the microprocessor technology that enables manufacturers to put an entire CPU on one chip. Businesses use personal computers for word processing, accounting, desktop publishing, and for running spreadsheet and database management applications. At home, the most popular use for personal computers is for playing games and computer-assisted learning.

Personal computers first appeared in the late 1970s. One of the first and most popular personal computers was the Apple II, introduced in 1977 by Apple Computer.

CONTENTS

Unit 1　History of Computer / 1

Unit 2　The Computer's Components / 21

Unit 3　Operating System / 37

Unit 4　Database System / 53

Unit 5　Programming Language / 67

Unit 6　Network and Internet / 81

Unit 7　Information Security / 95

Unit 8　Multimedia and Computer Games / 109

Unit 9　E-business / 123

Unit 10　Future of Computer / 135

Appendix A　Specialist English and Translation Skills / 151

Appendix B　Screen Prompt Message / 170

Appendix C　Index of Basic Vocabulary / 179

Appendix D　Concise Explanation of Common Abbreviations / 185

※ Unit 3　Operating System ※

What you have learned in this unit:

What you have learned in this unit:

Unit 4
Database System

After reading this unit and completing the exercises, you will be able to

☆ understand the phases of database system development in the computer history;

☆ be familiar with database function and approach of the relational database system;

☆ identify several different models of the database system;

☆ apply your knowledge when you store, access, share the computer's data and information.

Part I Lead-in

1. *Look at the pictures and have a discussion about them. Then produce the possible words and expressions that are related to the theme.*

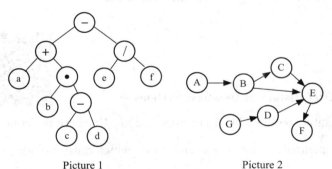

Picture 1   Picture 2   Picture 3

Write down the relevant words and expressions in the space provided below.

_____

_____

2. *Listen to the passage and then fill in the blanks with the exact words or phrases you have just heard.*

In broad themes, database system has historically been of three types: _____, _____, and object-oriented.

· 53 ·

Generally speaking, hierarchical databases arose in the 1960s, on the mainframe data processing technology of the time. Network databases are similar to hierarchical ones, but they allow multiple parent/child relations.In the 1970s, the rigorous mathematical work of Ted Codd and others created the now-ubiquitous _____. In the 1980s, largely because of the growing popularity of object-oriented programming, object databases gained a measure of popularity, especially to model so-called rich data, such as multimedia formats.

Part II Read and Understand

1. *Discuss the questions with your partners.*

1) What are the features of the Network Data model?

2) Where does the rich information of computer store?

3) What is the data and what is the database system?

4) How many types does the database system model have?

5) Talk about the difference between storing in a file and in database.

2. *Time to read it*.

↬ A Brief History of Database Systems ↫

Data are raw facts that constitute building blocks of information. Database is a collection of information and a means to manipulate data in a useful way, which must provide proper storage for large amounts of data, easy and fast access and facilitate the processing of data. Database Management System (DBMS) is a set of software that is used to define, store, manipulate and control the data in a database. From pre-stage flat-file system, to relational and object-relational systems, database technology has gone through several generations and its 40 years' history.

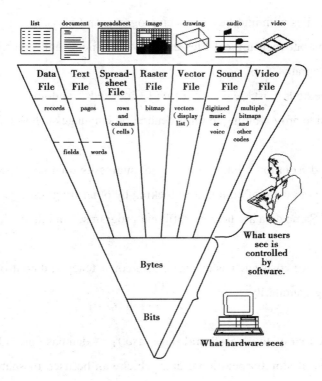

↬ The Evolution of the Database ↫

Ancient history: Data are not stored on disk; programmer defines both logical data structure and physical structure, such as storage structure, access methods, I/O modes, etc.

One data set per program: high data redundancy. There is no persistence. Random Access Memory (RAM) is expensive and limited, and programmer productivity is low.

1968 (File-based): predecessor of database. Data are maintained in a flat file. Processing characteristics determined by common use of magnetic tape medium.

* Data are stored in files with interface between programs and files.
* Mapping happens between logical files and physical files, and one file corresponds to one or several programs.
* Various access methods exist, e.g., sequential, indexed, random.
* It requires extensive programming in third-generation language such as COBOL, BASIC.
* Limitations:
  ☆ Separation and isolation: Each program maintains its own set of data, users of one Program may not aware of holding or blocking by other programs.
  ☆ Duplication: Same data are held by different programs, and thus, it wastes space and resources.
  ☆ High maintenance costs such as ensuring data consistency and controlling access.
  ☆ Coarse sharing granularity.
  ☆ Weak security.

1968—1980 (**Era of non-relational database**): A database provides integrated and structured collection of stored operational data which can be used or shared by application systems. Prominent hierarchical database model was IBM's first DBMS called IMS. Prominent network database model was CODASYL DBTG model; IDMS was the most popular network DBMS.

Hierarchical data model:

* In the mid-1960s, Rockwell worked with IBM to create Information Management System (IMS), and IMS DB/DC led the mainframe database market in 1970s and early 1980s.
* Based on binary trees; logically represented by an upside down tree, one-to-many relationship between parent and child records.
* Efficient searching; less redundant data; data independence; database security and integrity.

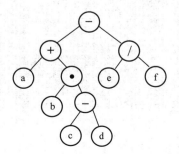

* Disadvantages:
  * ☆ Complex implementation.
  * ☆ Difficult to manage and short of standards, such as problems to add empty nodes and to handle many-to-many relationships.
  * ☆ Lack of structural independence.

Network data model:

* In early 1960s, Charles Bachmann developed first DBMS at Honeywell, Integrated Data Store (IDS).
* It was standardized in 1971 by the CODASYL group (Conference on Data Systems Languages).
* It directed acyclic graph with nodes and edges.

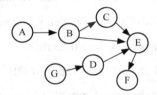

* It identified 3 database components: network schema—database organization; subschema—views of database per user; data management language—at low level and procedural.
* Each record can have multiple parents:
  * ☆ Composed of sets relationships, a set represents a one-to-many relationship between the owner and the member.
  * ☆ Each set has owner record and member record.
  * ☆ Member may have several owners.
* Main problems: System complexity and difficult to design and maintain; lack of structural independence.

> *The distinction of storing data in files and databases is that databases are intended to be used by multiple programs and types of users.*

1970—present (Era of relational database and database management system):
Based on relational calculus, shared collection of logically related data and a description of this data, they were designed to meet the information needs of an organization. System catalog/metadata provides description of data to enable program-data independence; logically related data comprises entities, attributes, and relationships of an organization's information. Data abstraction allows view level, a way of presenting data to a group of users and logical level, how data is understood to be when writing queries.

Hypothetical Relational Database Model

PubID	Publisher	PubAddress
03-4472822	Random House	123 4th Street, New York
04-7733903	Wiley and Sons	45 Lincoln Blvd, Chicago
03-4859223	O'Reilly Press	77 Boston Ave, Cambridge
03-3920886	City Lights Books	99 Market, San Francisco

AuthorID	AuthorName	AuthorBDay
345-28-2938	Haile Selassie	14-Aug-92
392-48-9965	Joe Blow	14-Mar-15
454-22-4012	Sally Hemmings	12-Sept-70
663-59-1254	Hannah Arendt	12-Mar-06

ISBN	AuthorID	PubID	Date	Title
1-34532-482-1	345-28-2938	03-4472822	1990	Cold Fusion for Dummies
1-38482-995-1	392-48-9965	04-7733903	1985	Macrame and Straw Tying
2-35921-499-4	454-22-4012	03-4859223	1952	Fluid Dynamics of Aquaducts
1-38278-293-4	663-59-1254	03-3920886	1967	Beads, Baskets & Revolution

* 1970: Ted Codd at IBM's San Jose Lab proposed relational models.

* Two major projects started and both were operational in late 1970s:

  ☆ INGRES at University of California: Berkeley became commercial and followed up POSTGRES which was incorporated into Informix.

  ☆ System R at IBM San Jose Lab later evolved into DB2, which became one of the first DBMS product based on the relational model. (Oracle produced a similar product just prior to DB2.)

* 1976: Peter Chen defined the Entity-Relationship (ER) model.

* 1980s: Maturation of the relational database technology. More relational based DBMSs were developed and SQL standard adopted by ISO and ANSI.

* 1985: Object-Oriented DBMS (OODBMS) develops. It was little success commercially because advantages did not justify the cost of converting billions of bytes of data to new format.

* 1990s: Incorporation of object-orientation in relational DBMS, new application areas, such as data warehousing and OLAP, web and Internet, Interest in text and multimedia, Enterprise Resource Planning (ERP) and Management Requirements Planning (MRP):

  ☆ 1991: Microsoft ships ACCESS, a personal DBMS created as element of Windows gradually supplanted all other personal DBMS products.

  ☆ 1995: First Internet database applications came out.

  ☆ 1997: XML applied to database processing, which solves long-standing database problems. Major vendors begin to integrate XML into DBMS products.

The 1990s, along with a rise in object-oriented programming, saw a growth in how data in various databases were handled. Programmers and designers began to treat the data in their databases as objects. That is to say that if a person's data were in a database, that person's attributes, such as their address, phone number, and age, were now considered to belong to that person instead of being extraneous data. This allows for relations between data to be relations to objects and their attributes and not to individual fields. The term "object-relational impedance mismatch" described the inconvenience of translating between programmed objects and database tables. Object databases and object-relational databases attempt to solve this problem by providing an object-oriented language (sometimes as extensions to SQL) that programmers can use as alternative to purely relational SQL. On the programming side, libraries known as object-relational mappings (ORMs) attempt to solve the same problem.

* 2000s: NoSQL and NewSQL.

XML databases are a type of structured document-oriented database that allows querying based on XML document attributes. XML databases are mostly used in enterprise database management, where XML is being used as the machine-to-machine data interoperability standard. XML database management systems include commercial software MarkLogic and Oracle Berkeley DB XML, and a free use software Clusterpoint Distributed XML/JSON Database. All are enterprise software database platforms and support industry standard ACID-compliant transaction processing with strong database consistency characteristics and high level of database security.

NoSQL databases are often very fast, do not require fixed table schemas, avoid join operations by storing denormalized data, and are designed to scale horizontally. The most popular NoSQL systems include MongoDB, Couchbase, Riak, Memcached, Redis, CouchDB, Hazelcast, Apache Cassandra, and HBase, which are all open-source software products.

In recent years, there was a high demand for massively distributed databases with high partition tolerance but according to the CAP theorem it is impossible for a distributed system to simultaneously provide consistency, availability, and partition tolerance guarantees. A distributed system can satisfy any two of these guarantees at the same time, but not all three. For that reason, many NoSQL databases are using what is called eventual consistency to provide both availability and partition tolerance guarantees with a reduced level of data consistency.

NewSQL is a class of modern relational databases that aims to provide the same scalable performance of NoSQL systems for online transaction processing (read-write) workloads while still using SQL and maintaining the ACID guarantees of a traditional database system. Such databases include Google F1/Spanner, CockroachDB, TiDB, ScaleBase, MemSQL,

NuoDB, and VoltDB.

The advanced database technology, along with Internet has proved faster communication and world-wide connectivity and ubiquitous publishing seems to lead information overload.

Words & Expressions

brief /bri:f/ adj. 简要的
raw /rɔ:/ adj. （材料，物质）自然状态的；未加工的；未经处理的
manipulate /mə'nɪpjuleɪt/ v. 操作；使用
facilitate /fə'sɪlɪteɪt/ v. 使便利；减轻……的困难
object-relational system 对象关系系统
go through 通过；复习
flat file 扁平文件
magnetic tape 磁带
sequential /sɪ'kwenʃl/ adj. 按次序的；顺序的；序列的
random /'rændəm/ adj. 随机的；随意的 n. 随意
isolation /ˌaɪsə'leɪʃn/ n. 隔离；孤立状态
duplication /ˌdju:plɪ'keɪʃn/ n. （不必要的）重复；复制
hierarchical /ˌhaɪə'rɑ:kɪkl/ adj. 按等级划分的
schema /'ski:mə/ n. 提要；纲要
view /vju:/ n. 视图
entity-relationship model 实体关系模型
（be）along with 跟随；伴随
object-oriented programming 面向对象编程
structured document-oriented 面向结构化文档的
transaction processing 事务处理

Notes

1. Database is a collection of information and a means to manipulate data in a useful way, which must provide proper storage for large amounts of data, easy and fast access and facilitate the processing of data. 数据库包括信息存储和数据处理两个方面，它需要提

供适合大量数据存储的空间，简单和快速的访问和方便的数据处理。

在这个句子中关系代词 which 引导的是非限定性定语从句，修饰限定 database。这种从句中关系代词和主语用逗号隔开，并且只有特殊语问词，如 which，who，where 等才能引导此类从句。

2. Difficult to manage and short of standards, such as problems to add empty nodes and to handle many-to-many relationships.（不足为）很难进行管理、缺乏标准，诸如增加空节点的问题及不容易处理多对多的关系。

在这个句子中，"Difficult to manage and short of standards"是省略句，省略了 it 和 is，后面句子是对前面的补充说明。

3. Composed of sets relationships, a set represents a one-to-many relationship between the owner and the member. 关系集的组成代表了主体和成员之间一对多的关系。

在这个句子中，"Composed of sets of relationships"为过去分词短语，表被动意义，其逻辑主语为 a set，相当于 "a set is composed of sets relationships"。

one-to-many 是一对多，相同的用法有：一对一（one-to-one）；多对多（many-to-many）。

4. I/O：输入输出。

5. CODASYL DBTG：数据系统语言会议数据库任务组。

6. IDMS：综合数据库管理系统。

7. IMS：信息管理系统，是 IBM 第一个层次性数据库管理系统。

8. ISO：国际标准化组织。

9. ANSI：美国国家标准学会。

10. ERP：企业资源计划。

11. MRP：物料需求计划。

12. XML：可扩展标记语言。

Part III Activities

1. *Review Questions.*

   Answer the following questions with information from the passage.

   1）Where were the computer information and data stored at the beginning?

   2）What are the disadvantages of ways the computer data are stored in a file?

   3）What are the ways of storing information in database?

   4）What is the primary function of relational DBMS?

5) In modern times, what are the manufacturers of the popular database management system product?

2. *Information to Complete.*

Complete the following sentences with the words and phrases from the passage.

Data are raw facts that compose construction blocks of information. Database is a set of information and a way to _____ data in a useful way, which must provide proper _____ for large amounts of data, easy and fast access and _____ _____ the processing of data. DBMS is a set of _____ that is used to define, store, manipulate and control the data in a database. From pre-stage flat-file system, to _____ system and network system, then to relational and object-_____ systems, database technology has gone through several generations and its 40 years history. The advanced database technology, along with Internet has proved faster _____ and world-wide connectivity, and ubiquitous publishing seems to lead information overload.

3. *Sentences to Translate.*

Translate the following sentences into English.

1) 存取数据是根据物理地址进行的，这种方式迫使程序员直接与物理设备打交道。
2) 信息是指经过加工处理后的数据，是被整理消化过的数据。
3) 数据是对客观事物特征的一种抽象的、符号化的表示。
4) 常见的库操作有四种：检索、插入、删除、更新。
5) 根据DBTG报告实现的系统一般称为DBTG系统。现有的网状数据库系统大都是采用DBTG方案的。
6) DBTG系统是典型的三级结构体系：子模式、模式、存储模式。

4. *Matching and Description.*

The passage you have read is about computers. The following are some useful expressions and explanations for computers. Read them carefully and find the items equivalent to those given in English in the table below. Then write the corresponding letter in the brackets.

1) a collection of information and a means to manipulate data in a useful way, which must provide proper storage for large amounts of data, easy and fast access and facilitate the

processing of data

2) a set of software that is used to define, store, manipulate and control the data in a database

3) a database model that views the real world as a set of basic objects (entities) and relationships among these objects (It is intended primarily for the DB design process by allowing the specification of an enterprise scheme. This represents the overall logical structure of the DB.)

4) a database that groups' data using common attributes found in the data set (The resulting "clumps" of organized data are much easier for people to understand. For example, a data set containing all the real estate transactions in a town can be grouped by the year the transaction occurred; or it can be grouped by the sale price of the transaction; or it can be grouped by the buyer's last name; and so on.)

5) a data model in which the data is organized into a tree-like structure (The structure allows repeating information using parent/child relationships: each parent can have many children but each child only has one parent. All attributes of a specific record are listed under an entity type.)

6) a data model conceived as a flexible way of representing objects and their relationships (The original inventor of network model was Charles Bachman, and it was developed into a standard specification published in 1969 by the CODASYL Consortium.)

a. 层次数据模型
b. 数据库管理系统
c. 实体关系模型
d. 关系型数据库
e. 数据库
f. 网络数据模型

( )( ) Database Management System (DBMS)
( )( ) relational database
( )( ) hierarchical data model
( )( ) network data mode
( )( ) Entity-Relationship (ER) model
( )( ) database

Part Ⅳ Further Reading

Big Data

Big data is a blanket term for any collection of data sets so large and complex that it becomes difficult to process using on-hand database management tools or traditional data processing applications.

The challenges include capture, curation, storage, search, sharing, transfer, analysis and visualization. The trend to larger data sets is due to the additional information derivable from analysis of a single large set of related data, as compared to separate smaller sets with the same total amount of data, allowing correlations to be found to spot business trends, determine quality of research, prevent diseases, link legal citations, combat crime, and determine real-time roadway traffic conditions.

A visualization of Wikipedia edits was created by IBM. At multiple terabytes in size, the text and images of Wikipedia are a classic example of big data.

As of 2012, limits on the size of data sets that are feasible to process in a reasonable amount of time were on the order of exabytes of data, that is, millions of terabytes. Scientists regularly encounter limitations due to large data sets in many areas, including meteorology, genomics, connectomics, complex physics simulations, and biological and environmental research. The limitations also affect Internet search, finance and business informatics. Data sets grow in size in part because they are increasingly being gathered by ubiquitous information-sensing mobile devices, aerial sensory technologies (remote sensing), software logs, cameras, microphones, Radio-Frequency Identification (RFID) readers, and wireless sensor networks. The world's technological per-capita capacity to store information has roughly doubled every 40 months since the 1980s; as of 2012, every day 2.5 exabytes of data were created. The challenge for large enterprises is determining who should own big data initiatives that straddle the entire organization.

Big data is difficult to work with using most relational database management systems and desktop statistics and visualization packages, requiring instead massively parallel software running on tens, hundreds, or even thousands of servers. What is considered "big data" varies depending on the capabilities of the organization managing the set, and on the capabilities of the applications that are traditionally used to process and analyze the data set in its domain. "For some organizations, facing hundreds of gigabytes of data for the first time may trigger a need to reconsider data management options. For others, it may take tens or hundreds of terabytes before data size becomes a significant consideration."

※ Unit 4 Database System ※

What you have learned in this unit:

What you have learned in this unit:

Unit 5
Programming Language

After reading this unit and completing the exercises, you will be able to

☆ understand what a programming language is;

☆ be familiar with the history of programming language;

☆ identify all the factors that should be considered when selecting a programming language;

☆ apply your knowledge when using programming language.

Part I Lead-in

1. *Look at the pictures and have a discussion about them. Then produce the possible words and expressions that are related to the theme.*

Picture 1

Picture 2   Picture 3

Write down the relevant words and expressions in the space provided below.

_____

_____

2. *Listen to the passage and then fill in the blanks with the exact words or phrases you have just heard.*

We have attempted to _____ a variety of _____

about the relative popularity of programming languages, mostly out of curiosity. To some _____ popularity does matter. However, it is _____ not the only thing to take into account when choosing a programming language. Most experienced _____ should be able to learn the basics of a new language in a week, and be productive with it in a _____ more weeks, although it will likely take much longer to truly master it.

Part II Read and Understand

1. *Discuss the questions with your partners.*

☆ Is JavaScript the future of programming?

☆ Where the first programming language firstly appeared?

☆ Did the invention of the computer precede the earliest programming languages?

☆ Talk about the categories of programming languages.

2. *Time to read it.*

↪ What Is a Programming Language ↩

A programming language is an artificial language designed to communicate instructions to a machine, particularly a computer. Programming languages can be used to create programs that control the behavior of a machine and/or to express algorithms.

The earliest programming languages preceded the invention of the computer, and were used to direct the behavior of machines such as Jacquard looms and player pianos. Thousands of different programming languages have been created, mainly in the computer field, and many more still are being created every year. Many programming languages require computation to be specified in an imperative form (i.e., as a sequence of operations to perform), while other languages utilize other forms of program specification such as the declarative form (i.e. the desired result is specified, not how to achieve it).

The description of a programming language is usually split into the two components of syntax (form) and semantics (meaning). Some languages are defined by a specification document (for example, the C programming language is specified by an ISO Standard), while other languages (such as Perl) have a dominant implementation that is treated as a reference.

→ History ←

NO Language, Just Wires

In 1946, the ENIAC was programmed by plugging wires from one socket to another. That led to the plugboards on tabulating machines and later to programming language.

Development of Low-level Languages

All computers operate by following machine language programs, a long sequence of instructions called machine code that is addressed to the hardware of the computer and is written in binary notation, which uses only the digits 1 and 0. First-generation languages, called machine languages, required the writing of long strings of binary numbers to represent such operations as "add", "subtract" "and compare". Later improvements allowed octal, decimal, or hexadecimal representation of the binary strings.

Because writing programs in machine language is impractical (it is tedious and error prone), symbolic, or assembly, languages—second-generation languages—were introduced in the early 1950s. They use simple mnemonics such as A for "add" or M for "multiply", which are translated into machine language by a computer program called an assembler. The assembler then turns that program into a machine language program. An extension of such a language is the macro instruction, a mnemonic (such as "READ") for which the assembler substitutes a series of simpler mnemonics. The resulting machine language programs, however, are specific to one type of computer and will usually not run on a computer with a different type of Central Processing Unit (CPU).

Evolution of High-level Languages

The lack of portability between different computers led to the development of high-level languages—so called because they permitted a programmer to ignore many low-level details of the computer's hardware. Further, it was recognized that the closer the syntax, rules, and mnemonics of the programming language could be to "natural language" the less likely it became that the programmer would inadvertently introduce errors (called "bugs") into the program. Hence, in the mid-1950s a third generation of languages came into use. These algorithmic, or procedural, languages are designed for solving a particular type of problem. Unlike machine or symbolic languages, they vary little between computers. They must be translated into machine code by a program called a compiler or interpreter.

Early computers were used almost exclusively by scientists, and the first high-level language, Fortran (Formula Translation), was developed (1953—1957) for scientific and engineering applications by John Backus at the IBM Corp. A program that handled recursive algorithms better, LISP (List Processing), was developed by John McCarthy at the Massachusetts Institute of Technology in the early 1950s; implemented in 1959, it has become the standard language for the artificial intelligence community. COBOL (Common Business Oriented Language), the first language intended for commercial applications, is still widely used; it was developed by a committee of computer manufacturers and users under the leadership of Grace Hopper, a U.S. Navy programmer, in 1959. ALGOL (Algorithmic Language), developed in Europe about 1958, is used primarily in mathematics and science, as is APL (A Programming Language), published in the United States in 1962 by Kenneth Iverson. PL/1 (Programming Language 1), developed in the late 1960s by the IBM Corp., and ADA (for Ada Lovelace), developed in 1981 by the U.S. Dept. of Defense, are designed for both business and scientific use.

BASIC (Beginner's All-purpose Symbolic Instruction Code) was developed by two Dartmouth College professors, John Kemeny and Thomas Kurtz, as a teaching tool for undergraduates (1966); it subsequently became the primary language of the personal computer revolution. In 1971, Swiss professor Nicholas Wirth developed a more structured language for teaching that he named Pascal (for French mathematician Blaise Pascal, who built the first successful mechanical calculator). Modula 2, a Pascallike language for commercial and mathematical applications, was introduced by Wirth in 1982. Ten years before that, to implement the UNIX operating system, Dennis Ritchie of Bell Laboratories produced a language that he called C; along with its extensions, called C++, developed by Bjarne Stroustrup of Bell Laboratories, it has perhaps become the most widely used general-purpose language among professional programmers because of its ability to deal with the rigors of object-oriented programming. Java is an object-oriented language similar to C++ but simplified to eliminate features that are prone to programming errors. Java was developed specifically as a network-oriented language, for writing programs that can be safely downloaded through the Internet and immediately run without fear of computer viruses. Using small Java programs called applets, World Wide Web pages can be developed that include a full range of multimedia functions.

Fourth-generation languages are nonprocedural—they specify what is to be accomplished without describing how. The first one, FORTH, developed in 1970 by American astronomer Charles Moore, is used in scientific and industrial control applications. Most fourth-generation languages are written for specific purposes. Fifth-generation languages, which are still in their infancy, are an outgrowth of artificial intelligence research. PROLOG (Programming Logic), developed by French computer scientist Alain Colmerauer and logician Philippe Roussel in the early 1970s, is useful for programming logical processes and making deductions automatically.

Many other languages have been designed to meet specialized needs. GPSS (General Purpose System Simulator) is used for modeling physical and environmental events, and SNOBOL (String-Oriented Symbolic Language) is designed for pattern matching and list processing. LOGO, a version of LISP, was developed in the 1960s to help children learn about computers. PILOT (Programmed Instruction Learning, Or Testing) is used in writing instructional software, and Occam is a nonsequential language that optimizes the execution of a program's instructions in parallel-processing systems.

There are also procedural languages that operate solely within a larger program to customize it to a user's particular needs. These include the programming languages of several database and statistical programs, the scripting languages of communications programs, and the macro languages of word-processing programs.

↝ 10 Programming Languages You Should Learn in 2020 ↜

The tech sector is booming. If you've used a smartphone or logged on to a computer at least once in the last few years, you've probably noticed this.

As a result, coding skills are in high demand, with programming jobs paying significantly more than the average position. Even beyond the tech world, an understanding of at least one programming language makes an impressive addition to any resume.

Java

Java is a class-based, Object-Oriented Programming (OOP) language developed by Sun Microsystems in the 1990s. It's one of the most in-demand programming languages, a standard for enterprise software, web-based content, games and mobile apps, as well as the Android operating system. Java is designed to work across multiple software platforms, meaning a program written on Mac OS X, for example, could also run on Windows.

C Language

A general-purpose, imperative programming language developed in the early 1970s, C is the oldest and most widely used language, providing the building blocks for other popular languages, such as C#, Java, JavaScript and Python. C is mostly used for implementing operating systems and embedded applications.

Because it provides the foundation for many other languages, it is advisable to learn C (and C++) before moving on to others.

C++

C++ powers major software like Firefox, Winamp and Adobe programs. It's used to develop systems software, application software, high-performance server and client applications and video games.

C#

Pronounced "C-sharp", C# is a multi-paradigm language developed by Microsoft as part of its NET initiative. Combining principles from C and C++, C# is a general-purpose language used to develop software for Microsoft and Windows platforms.

Objective-C

Objective-C is a general-purpose, object-oriented programming language used by the Apple operating system. It powers Apple's OS X and iOS, as well as its APIs, and can be used to create iPhone apps, which has generated a huge demand for this once-outmoded programming language.

PHP

PHP (Professional Hypertext Processor) is a free, server-side scripting language designed for dynamic websites and app development. It can be directly embedded into an HTML source document rather than an external file, which has made it a popular programming language for web developers. PHP powers more than 200 million websites, including Wordpress, Digg and Facebook.

Python

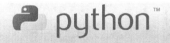

Python is a high-level, server-side scripting language for websites and mobile apps. It's considered a fairly easy language for beginners due to its readability and compact syntax, meaning developers can use fewer lines of code to express a concept than they would in other languages. It powers the web apps for Instagram, Pinterest and Rdio through its associated web framework, Django, and is used by Google, Yahoo! and NASA.

Ruby

A dynamic, object-oriented scripting language for developing websites and mobile apps, Ruby is designed to be simple and easy to write. It powers the Ruby on Rails (or Rails) framework, which is used on Scribd, GitHub, Groupon and Shopify. Like Python, Ruby is considered a fairly user-friendly language for beginners.

JavaScript

JavaScript is a client and server-side scripting language developed by Netscape that derives much of its syntax from C. It can be used across multiple web browsers and is considered essential for developing interactive or animated web functions. It is also used in game development and writing desktop applications. JavaScript interpreters are embedded in Google's Chrome extensions, Apple's Safari extensions, Adobe Acrobat and Reader, and Adobe's Creative Suite.

SQL

SQL (Structured Query Language) is a special-purpose language for managing data in relational database management systems. It is most commonly used for its "Query" function, which searches informational databases. SQL was standardized by the American National Standards Institute (ANSI) and the International Organization for Standardization (ISO) in the 1980s.

Words & Expressions

programming language　编程语言
artificial language　人工语言
player piano　自动钢琴
program specification　程序规范；程序规格说明
declarative form　声明形式
syntax and semantics　语法和语义
machine code　机器码
high-level　高级
scripting language　脚本语言
class-based　基于类的
imperative programming language　命令式编程语言
multi-paradigm language　多范式语言
combine /kəmˈbaɪn/ v.　结合
server-side　服务器端
suite /swiːt/ n.　套装软件

Notes

1. instruction：指令。"指令"是由指令集架构定义的单个 CPU 操作。在更广泛的意义上，"指令"可以是任何可执行程序的元素的表述，例如字节码。

2. machine code：机器码。机器语言（machine language）是一种指令集的体系。这种指令集，被称为机器码（machine code），是电脑的 CPU 可直接解读的数据。机器码有时也被称为原生码（native code），这个名词比较强调某种编程语言或库，它与运行平台相关。

3. OOP：面向对象编程是一种计算机编程架构。OOP 的一条基本原则是计算机程序是由单个能够起到子程序作用的单元或对象组合而成。OOP 达到了软件工程的三个主要目标：重用性、灵活性和扩展性。为了实现整体运算，每个对象都能够接收信息、处理数据和向其他对象发送信息。

Part III Activities

1. *Review Questions.*

Answer the following questions with information from the passage.

1）What is a programming language?

2）How many programming languages are introduced in the article?

3）What are the programming languages that are commonly used today?

4）What is object-oriented programming language?

2. *Information to Complete.*

1）The description of a programming language is usually split into the two components of _____ （form）and _____ （meaning）.

2）Java is a _____, object-oriented programming language developed by Sun Microsystems in the 1990s.

3）Python is a high-level, server-side _____ language for websites and mobile apps.

3. *Sentences to Translate.*

Translate the following sentences into English.

1）编程语言的描述通常分成两个部分：语法（形式）和语义（意义）。

2）许多其他语言的设计都是为了满足特殊的需求。

3）Java 是一种类似于 C++ 的面向对象语言，但是做了简化以消除容易出现编程错误的功能。

4）C++结合了面向对象和系统编程。

5）Java后来被用于服务器端编程。

4. *Matching and Description.*

The passage you have read is about computers. The following are some useful expressions and explanations for computers. Read them carefully and find the items equivalent to those given in English in the table below. Then write the corresponding letter in the brackets.

1）an elemental language of computers, consisting of a stream of 0's and 1's

2）a low-level programming language for a computer, or other programmable device, in which there is a very strong (generally one-to-one) correspondence between the language and the architecture's machine code instructions

3）a programming language model organized around objects rather than "actions" and data rather than logic

4）a sequence of instructions, written to perform a specified task with a computer

5）an order given to a computer processor by a computer program

a. 面向对象编程　　　　　　　　（　）（　）program
b. 汇编语言　　　　　　　　　　（　）（　）instruction
c. 机器码　　　　　　　　　　　（　）（　）OOP
d. 指令　　　　　　　　　　　　（　）（　）assembly language
e. 程序　　　　　　　　　　　　（　）（　）machine code

Part Ⅳ Further Reading

Formal Language

In mathematics, computer science, and linguistics, a formal language is a set of strings of symbols that may be constrained by rules that are specific to it.

The alphabet of a formal language is the set of symbols, letters, or tokens from which the strings of the language may be formed; frequently it is required to be finite. The strings formed from this alphabet are called words, and the words that belong to a particular formal language are sometimes called well-formed words or well-formed formulas. A formal language is often defined by means of a formal grammar such as a regular grammar or context-free grammar, also called its formation rule.

The field of formal language theory studies primarily the purely syntactical aspects of such languages—that is, their internal structural patterns. Formal language theory sprang out of linguistics, as a way of understanding the syntactic regularities of natural languages. In computer science, formal languages are used among others as the basis for defining the grammar of programming languages and formalized versions of subsets of natural languages in which the words of the language represent concepts that are associated with particular meanings or semantics. In computational complexity theory, decision problems are typically defined as formal languages, and complexity classes are defined as the sets of the formal languages that can be parsed by machines with limited computational power. In logic and the foundations of mathematics, formal languages are used to represent the syntax of axiomatic systems, and mathematical formalism is the philosophy that all of mathematics can be reduced to the syntactic manipulation of formal languages in this way.

History

The first formal language is thought to be the one used by Gottlob Frege in his Begriffsschrift (1879), literally meaning "concept writing", and which Frege described as a "formal language of pure thought".

Axel Thue's early Semi-Thue system which can be used for rewriting strings was influential on formal grammars.

Definition

A formal language L over an alphabet Σ is a subset of Σ^*, which is a set of words over that alphabet. Sometimes the sets of words are grouped into expressions, whereas rules and constraints may be formulated for the creation of "well-formed expressions".

In computer science and mathematics, which do not usually deal with natural languages, the adjective "formal" is often omitted as redundant.

While formal language theory usually concerns itself with formal languages that are described by some syntactical rules, the actual definition of the concept "formal language" is only as above: a (possibly infinite) set of finite-length strings composed from a given alphabet, no more nor less. In practice, there are many languages that can be described by rules, such as regular languages or context-free languages. The notion of a formal grammar may be closer to the intuitive concept of a "language", one described by syntactic rules. By an abuse of the definition, a particular formal language is often thought of as being equipped with a formal grammar that describes it.

What you have learned in this unit:

What you have learned in this unit:

Unit 6
Network and Internet

After reading this unit and completing the exercises, you will be able to
☆ understand the different kinds of network;
☆ be familiar with the common network components;
☆ describe what is the difference between Network and Internet;
☆ list the common applications of Internet.

Part I   Lead-in

1. *Look at the pictures and have a discussion about them. Then produce the possible words and expressions that are related to the theme.*

 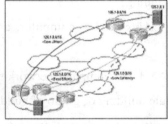 

　　　　Picture 1　　　　　　　　Picture 2　　　　　　　　Picture 3

Write down the relevant words and expressions in the space provided below.

_____

_____

2. *Listen to the passage and then fill in the blanks with the exact words or phrases you have just heard.*

　　Most LANs connect workstations and _____. Each node (individual computer) in a _____ has its own CPU with which it executes _____, but it also is able to access _____ and devices anywhere on the LAN. This means that many users can share devices, such as laser printers,

· 81 ·

as well as data. Users can also use the LAN to _____ with each other, by sending e-mail or engaging in chat sessions.

Part II Read and Understand

1. Discuss the questions with your partners.

☆ What is a computer network?

☆ When did the first computer network appear?

☆ What can we do with Internet?

☆ Is the Internet same to World Wide Web? If no, what is the difference between them?

2. Time to read it.

↦ **What Is Network** ↤

A network is a group of two or more computer systems linked together. There are many types of computer networks, including:

Local-Area Networks (LANs): The computers are geographically close together (that is, in the same building).

A Local-Area Network (LAN) is a computer network that spans a relatively small area. Most LANs are confined to a single building or group of buildings.

LANs are capable of transmitting data at very fast rates, much faster than data can be transmitted over a telephone line; but the distances are limited, and there is also a limit on the number of computers that can be attached to a single LAN.

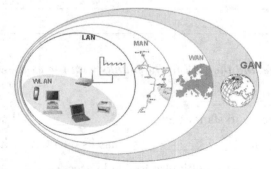

Wide-Area Networks (WANs): The computers are farther apart and are connected by telephone lines or radio waves.

A computer network that spans a relatively large geographical area. Typically, a WAN consists of two or more Local-Area Networks (LANs).

Computers connected to a wide-area network are often connected through public networks, such as the telephone system. They can also be connected through leased lines or satellites. The largest WAN in existence is the Internet.

***Metropolitan-Area Networks* (*MANs*)**: A data network is designed for a town or city.

In terms of geographic breadth, MANs are larger than Local-Area Networks (LANs), but smaller than Wide-Area Networks (WANs). MANs are usually characterized by very high-speed connections using fiber optical cable or other digital media.

The Other Categories of Networks

In addition to these types, the following characteristics are also used to categorize different types of networks:

Topology: It refers of the geometric arrangement of a computer system. Common topologies include a bus, a star, and a ring.

Protocol: The protocol defines a common set of rules and signals that computers on the network use to communicate. One of the most popular protocols for LANs is called Ethernet. Another popular LAN protocol for PCs is the IBM token-ring network.

Architecture: Networks can be broadly classified as using either a peer-to-peer or client/server architecture. Computers on a network are sometimes called nodes. Computers and devices that allocate resources for a network are called servers.

↪ Internet ↩

The Internet is a global system of interconnected computer networks that use the standard Internet protocol suite (TCP/IP) to link several billion devices worldwide. It is a network of networks that consists of millions of private, public, academic, business, and government networks, of local to global scope, that are linked by a broad array of electronic, wireless, and optical networking technologies. The Internet carries an extensive range of information resources and services, such as the inter-

linked hypertext documents and applications of the World Wide Web (WWW), the infrastructure to support email, and peer-to-peer networks for file sharing and telephony.

Protocol

In information technology, a protocol is the special set of rules that end points in a telecommunication connection use when they communicate. Protocols specify interactions between the communicating entities.

Protocols exist at several levels in a telecommunication connection. For example, there are protocols for the data interchange at the hardware device level and protocols for data interchange at the application program level. In the standard model known as Open Systems Interconnection (OSI), there are one or more protocols at each layer in the telecommunication exchange that both ends of the exchange must recognize and observe. Protocols are often described in an industry or international standard.

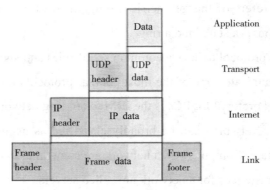

The TCP/IP Internet protocols, a common example, consist of:

☆ Transmission Control Protocol (TCP), which uses a set of rules to exchange messages with other Internet points at the information packet level.

☆ Internet Protocol (IP), which uses a set of rules to send and receive messages at the Internet address level.

☆ Additional protocols include the Hypertext Transfer Protocol (HTTP) and File Transfer Protocol (FTP), each with defined sets of rules to use with corresponding programs elsewhere on the Internet.

There are many other Internet protocols, such as the Border Gateway Protocol (BGP) and the Dynamic Host Configuration Protocol (DHCP).

Routing

Internet service providers connect customers, which represent the bottom of the routing hierarchy, to customers of other ISPs via other higher or same-tier networks. At the top of the routing hierarchy are the tier 1 networks, large telecommunication companies that exchange traffic directly with each other via peering agreements. Tier 2 and lower level networks buy Internet transit from other providers to reach at least some parties on the global Internet, though they may also engage in peering. An ISP may use a single upstream provider for connectivity, or implement multi homing to achieve redundancy and load balancing. Internet exchange points are major traffic exchanges with physical connections to multiple ISPs.

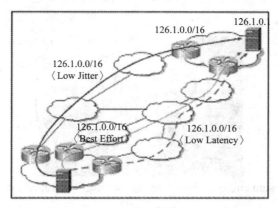

Computers and routers use a routing table in their operating system to direct IP packets to the next-hop router or destination. Routing tables are maintained by manual configuration or automatically by routing protocols. End-nodes typically use a default route that points toward an ISP providing transit, while ISP routers use the Border Gateway Protocol to establish the most efficient routing across the complex connections of the global Internet.

Large organizations, such as academic institutions, large enterprises, and governments, may perform the same function as ISPs, engaging in peering and purchasing transit on behalf of their internal networks. Research networks tend to interconnect with large sub networks such as GEANT, GLORIAD, Internet 2, and the UK's national research and education network, JANET.

Modern Uses
- Communication

At the moment the easiest thing that can be done using the Internet is that we can communicate with the people living far away from us with extreme ease. Earlier the communication used to be a daunting task but all that chanced once the Internet came into the life of the common people. Now people can not only chat but can also do the video conferencing. Communication is the most important gift that the Internet has given to the common man.

- Research

In order to do research, you need to go through hundreds of books as well as the references and that was one of the most difficult jobs to do earlier. Since the Internet came into life, everything is available just a click away. You just have to search for the concerned topic and you will get hundreds of references that may be beneficial for your research.

- Financial Transaction

With the use of the Internet in the financial transaction, your work has become a lot easier. Now you don't need to stand in the queue at the branch of your particular bank rather you can just log in on to the bank website with the credential that has been provided to you by the bank and then can do any transaction related to finance at your will. With the ability to do the financial transaction easily over the Internet you can purchase or sell items so easily. Financial transaction can be considered as one of the best uses of resource in the right direction.

- Leisure

Leisure is the option that we have next in the list. Right from watching your favorite videos to listening songs, watching movies, playing games, chatting with the loved ones has been possible due to the Internet. Internet has progressed with so much pace that today whenever you get time, you just move on to the Internet and enjoy such activities which help you to relax. Leisure is one of the most important uses of the Internet and it surely has one thing that attracts people towards it.

- Shopping

Shopping has now become one of the most pleasing things to do using the Internet. Whenever you find time, just visit the concerned websites and order the items that you need from there. Those items will be delivered to you in best possible time. There is huge number of options available for a common people to buy or to sell any particular item using the Internet. Using the Internet now make it possible to buy products from all over the world.

Words & Expressions

computer network 计算机网络
geographical /ˌdʒiːəˈɡræfɪkl/ *adj*. 地理的
leased line 专线
fiber optical cable 光纤

topology /tə'pɒlədʒi/ n. 拓扑结构
token-ring network 令牌环网
peer-to-peer 对等网
client/server architecture 客户端/服务器体系结构
Internet protocol suite 互联网协议族
hypertext /'haɪpətekst/ n. 超文本
Border Gateway Protocol 边界网关协议
routing table 路由表
default route 默认路由
financial /faɪ'nænʃl/ adj. 金融的；财务的

Notes

1. In information technology, a protocol is the special set of rules that end points in a telecommunication connection use when they communicate. 在信息技术中，协议是通信的时候应用在通信连接结点上的一组规则。

   that 此处引导定语从句，修饰 rules。

2. At the moment the easiest thing that can be done using the Internet is that we can communicate with the people living far away from us with extreme ease. 目前，可以使用互联网来完成最简单的事情是：我们与生活在远处的人沟通变得唾手可得。

   with extreme ease 意为"唾手可得"。

3. TCP/IP：Transmission Control Protocol/Internet Protocol，传输控制协议/因特网互联协议，又名网络通信协议，是 Internet 最基本的协议、Internet 国际互联网络的基础，由网络层的 IP 协议和传输层的 TCP 协议组成。

4. WWW：World Wide Web，中文名字为"万维网""环球网"等，常简称为 Web。分为 Web 客户端和 Web 服务器程序。WWW 可以让 Web 客户端（常用浏览器）访问浏览 Web 服务器上的页面。

5. HTTP：Hypertext Transfer Protocol，超文本传输协议，是一种详细规定了浏览器和万维网服务器之间互相通信的规则，通过因特网传送万维网文档的数据传送协议。

Part III Activities

1. *Review Questions*.

Answer the following questions with information from the passage.

1）What are the types of networks?

2）What is the difference between LAN and WAN?

3）Please generally discuss about how the computer evolved from mechanical devices to electronic digital devices.

4）How do the computers find each other in the Internet?

5）Please list the common uses of the Internet.

2. *Information to Complete*.

Complete the following sentences with the words and phrases from the passage.

1）A Local-Area Network（LAN）is a _____ that spans a relatively small area. Most _____ are confined to a single _____ or group of buildings.

2）Computers on a _____ are sometimes called nodes. Computers and _____ _____ that allocate resources for _____ are called servers.

3）Protocols exist at several _____ in a telecommunication _____ _____.

3. *Sentences to Translate*.

Translate the following sentences into English.

1）局域网（LAN)是指跨越一个相对较小的区域的计算机网络。

2）现有的最大广域网是互联网。

3）常见的拓扑结构包括总线型、星型和环型。

4）路由表由手工配置或者路由协议自动维护。

5）通信是互联网给人们最重要的礼物。

4. *Matching and Description*.

The passage you have read is about computers. The following are some useful expressions and explanations for computers. Read them carefully and find the items equivalent to those given in

English in the table below. Then write the corresponding letter in the brackets.

1) an international, archival journal providing a publication vehicle for complete coverage of all topics of the interest to those involved in the computer communications networking area

2) a cable containing one or more optical fibers that are used to carry light

3) the arrangement of the various elements (links, nodes, etc.) of a computer network

4) text displayed on a computer display or other electronic devices with references (hyperlinks) to other text which the reader can immediately access, or where text can be revealed progressively at multiple levels of detail

5) a set of rules, often viewed in table format, that is used to determine where data packets traveling over an Internet Protocol (IP) network will be directed

6) the special set of rules that end points in a telecommunication connection use when they communicate

a. 超文本
b. 路由表
c. 计算机网络
d. 光纤
e. 协议
f. 网络拓扑

( ) ( ) computer network
( ) ( ) fiber optical cable
( ) ( ) network topology
( ) ( ) hypertext
( ) ( ) routing table
( ) ( ) protocol

Part IV Further Reading

Internet of Things

The Internet of Things (IoT) refers to uniquely identifiable objects and their virtual representations in an Internet-like structure. The term Internet of Things was proposed by Kevin Ashton in 1999 though the concept has been discussed since at least 1991. The concept of the Internet of Things first became popular through the Auto-ID Center at MIT and related market analysis publications. Radio-Frequency Identification (RFID) was seen as a prerequisite for the Internet of Things in the early days. If all objects and people in daily life were equipped with identifiers, they could be managed and inventoried by computers. Besides using RFID, the tagging of things may be achieved through such technologies as near field communication, barcodes, Quick Response (QR) codes and digital watermarking.

Equipping all objects in the world with minuscule identifying devices or machine-readable identifiers could transform daily life. For instance, business may no longer run out of stock or generate waste products, as involved parties would know which products are required and consumed. A person's ability to interact with objects could be altered remotely based on immediate or present needs, in accordance with existing end-user agreements. For example, such technology could enable much more powerful control of content creators and owners over their creations by better applying copyright restrictions and digital restrictions management, so a customer buying a Blu-ray disc containing a movie could choose to pay a high price and be able to watch the movie for a whole year, pay a moderate price and have the right to watch the movie for a week, or pay a low fee every time she or he watches the movie.

Today however, the term Internet of Things (commonly abbreviated as IoT) is used to denote advanced connectivity of devices, systems and services that goes beyond the traditional M2M and covers a variety of protocols, domains and applications.

According to Gartner, there will be nearly 26 billion devices on the Internet of Things by 2020. According to ABI Research, more than 30 billion devices will be wirelessly connected to the Internet of Things (Internet of Everything) by 2020. Cisco created a dynamic "connections counter" to track the estimated number of connected things from July 2013 until July 2020 (methodology included). This concept, where devices connect to the Internet/Web via low-power radio, is the most active research area in IoT. The low-power radios do not need to use Wi-Fi or Bluetooth. Lower-power and lower-cost alternatives are being explored under the category of Chirp Networks.

All of the sensors and machine-readable identifiers needed to make the Internet of Things a reality will have to use IPv6 to accommodate the extremely large address space required. Even if the supply of IPv4 addresses were not to be exhausted soon, the size of IPv4 itself is not large enough to support the Internet of Things. To a large extent, the Internet of Things will be the ultimate driver of global adoption of IPv6 in the coming years.

※ Unit 6　Network and Internet ※

What you have learned in this unit:

What you have learned in this unit:

Unit 7
Information Security

After reading this unit and completing the exercises, you will be able to

☆ understand the definition and key characteristics of information security;
☆ be familiar with the history of information security;
☆ use the knowledge of information security to protect the computer;
☆ identify the importance of information technology.

Part Ⅰ Lead-in

1. Look at the pictures and have a discussion about them. Then produce the possible words and expressions that are related to the theme.

Picture 1

Picture 2

Picture 3

Picture 4

Write down the relevant words and expressions in the space provided below.

_____

_____

2. Listen to the passage and then fill in the blanks with the exact words or phrases you have just heard.

Encoding _____ is one of the oldest ways of securing written information. Governments and military organizations often use encryption _____ to ensure that secret messages will be unreadable if they are intercepted by the wrong person. Encryption methods can include simple substitution _____, like switching each letter for a corresponding _____, or more complex systems that require complicated algorithms for decryption. As long as the code method is kept secret, encryption can be a good basic method of _____ security.

Part II Read and Understand

1. Discuss the questions with your partners.

☆ Does a computer necessarily mean a home desktop?
☆ Discuss the privacy issues related to the Internet.
☆ Describe the security threats posed by hacker.
☆ Discuss ways that individuals and organizations protect their security.
☆ Talk about when the Caesar cipher appear.

2. Time to read it.

↪ **Information Security** ↩

Information security, sometimes shortened to InfoSec, is the practice of defending information from unauthorized access, use, disclosure, disruption, modification, perusal, inspection, recording or destruction. It is a general term that can be used regardless of the form the data may take (electronic, physical, etc.).

Two Major Aspects of Information Security:

- **IT Security**

Sometimes referred to as computer security, information technology security is information security applied to technology (most often some form of computer system). It is worthwhile to note that a computer does not necessarily mean a home desktop. A computer is any device with a processor and some memory. Such devices can range from non-networked standalone devices as simple as calculators, to networked mobile computing devices such as smartphones and tablet computers.

- **Information Assurance**

The act of ensuring that data is not lost when critical issues arise. These issues include but are not limited to: natural disasters, computer/server malfunction, physical theft, or any other instance where data has the potential of being lost. Since most information is stored on computers in our modern era, information assurance is typically dealt with by IT security specialists. One of the most common methods of providing information assurance is to have an off-site backup of the data in case one of the mentioned issues arise.

Governments, military, corporations, financial institutions, hospitals and private businesses amass a great deal of confidential information about their employees, customers, products, research and financial status. Most of this information is now collected, processed and stored on electronic computers and transmitted across networks to other computers.

Should confidential information about a business' customers or finances or new product line fall into the hands of a competitor or a black hat hacker, a business and its customers could suffer widespread, irreparable financial loss, not to mention damage to the company's reputation. Protecting confidential information is a business requirement and in many cases also an ethical and legal requirement.

For the individual, information security has a significant effect on privacy, which is viewed very differently in different cultures. The field of information security has grown and evolved significantly in recent years. There are many ways of gaining entry into the field as a career. It offers many areas for specialization including securing network(s) and allied infrastructure, securing applications and databases, security testing, information systems auditing, business continuity planning and digital forensics, etc.

History

Julius Caesar is credited with the invention of the Caesar cipher c. 50 B. C. , which was created in order to prevent his secret messages from being read and a message fall into the wrong hands. Sensitive information was marked up to indicate that it should be protected and transported by trusted persons, guarded and stored in a secure environment or strong box. As postal services expanded, governments created official organizations to intercept, decipher, read and reseal letters (e.g. the UK Secret Office and Deciphering Branch in 1653).

In the mid-19th century more complex classification systems were developed to allow governments to manage their information according to the degree of sensitivity. The British Government codified this, to some extent, with the publication of the *Official Secrets Act* in 1889. By the time of the First World War, multi-tier classification systems were used to communicate information to and from various fronts, which encouraged greater use of code making and breaking sections in diplomatic and military headquarters. In the United Kingdom this led to the creation of the Government Code and Cypher School in 1919. Encoding became more sophisticated between the wars as machines were employed to scramble and unscramble information. The volume of information shared by the Allied countries during the Second World War necessitated formal alignment of classification systems and procedural controls. An arcane range of markings evolved to indicate who could handle documents (usually officers rather than men) and where they should be stored as increasingly complex safes and storage facilities were developed. Procedures evolved to ensure documents were destroyed properly and it was the failure to follow these procedures which led to some of the greatest intelligence coups of the war (e. g. U-570).

The end of the 20th century and early years of the 21st century saw rapid advancements in telecommunications, computing hardware and software, and data encryption. The availability of smaller, more powerful and less expensive computing equipment made electronic data processing within the reach of small business and the home user. These computers quickly became interconnected through the Internet.

The rapid growth and widespread use of electronic data processing and electronic business conducted through the Internet, along with numerous occurrences of international terrorism, fueled the need for better methods of protecting the computers and the information they store, process and transmit. The academic disciplines of computer security and information assurance emerged along with numerous professional organizations—all sharing the common goals of ensuring the security and reliability of information systems.

↪ The Ways to Protect Information Security ↩

Information security is the process of protecting the availability, privacy, and integrity of data. While the term often describes measures and methods of increasing computer security, it also refers to the protection of any type of important data, such as personal diaries or the classified plot details of an upcoming book. No security system is foolproof, but taking basic and practical steps to protect data is critical for good information security.

Password Protection

Using passwords is one of the most basic methods of improving information security. This measure reduces the number of people who have easy access to the information, since only those with approved codes can reach it. Unfortunately, passwords are not foolproof, and hacking programs can run through millions of possible codes in just seconds. Passwords can also be breached through carelessness, such as by leaving a public computer logged into an account or using a too simple code, like "password" or "1234".

Antivirus and Malware Protection

One way that hackers gain access to secure information is through malware, which includes computer viruses, spyware, worms, and other programs. These pieces of code are installed on computers to steal information, limit usability, record user actions, or destroy data. Using strong antivirus software is one of the best ways of improving information security. Antivirus programs scan the system to check for any known malicious software, and most will warn the user if he or she is on a webpage that contains a potential virus. Most programs will also perform a scan of the entire system on command, identifying and destroying any harmful objects.

Firewalls

A firewall helps maintain computer information security by preventing unauthorized access to a network. There are several ways to do this, including by limiting the types of data allowed in and out of the network, re-routing network information through a proxy server to hide the real address of the computer, or by monitoring the characteristics of the data to determine if it's trustworthy. In essence, firewalls filter the information that passes through them, only allowing authorized content in. Specific websites, protocols (like File Transfer Protocol or FTP), and even words can be blocked from coming in, as can outside access to computers within the firewall.

Legal Liability

Businesses and industries can also maintain information security by using privacy laws. Workers at a company that handles secure data may be required to sign Non-Disclosure Agreements (NDAs), which forbid them from revealing or discussing any classified topics. If an employee attempts to give or sell secrets to a competitor or other unapproved source, the company can use the NDA as grounds for legal proceedings. The use of liability laws can help companies preserve their trademarks, internal processes, and research with some degree of reliability.

Training and Common Sense

One of the greatest dangers to computer data security is human error or ignorance. Those responsible for using or running a computer network must be carefully trained in order to avoid accidentally opening the system to hackers. In the workplace, creating a training program that includes information on existing security measures as well as permitted and prohibited computer usage can reduce breaches in internal security. Family members on a home network should be taught about running virus scans, identifying potential Internet threats, and protecting personal information online.

Words & Expressions

unauthorized access　非授权存取
ensure /ɪnˈʃʊə/ v.　保证；确保
critical issue　关键问题
privacy /ˈprɪvəsɪ/ n.　隐私；不受公众干扰的状态
security testing　安全性测试
digital forensics　数字取证
Caesar cipher　恺撒密码
Official Secrets Act　《官方保密法》
data encryption　数据加密
electronic data processing　电子数据处理
integrity /ɪnˈtegrətɪ/ n.　完整性
(be) not foolproof　不是万无一失
hacking program　黑客程序
virus /ˈvaɪrəs/ n.　病毒
spyware /ˈspaɪweə/ n.　间谍软件；间谍程序
worm /wɜːm/ n.　蠕虫
antivirus program　防病毒程序
potential /pəˈtenʃl/ adj.　潜在的
firewall /ˈfaɪəwɔːl/ n.　防火墙
proxy server　代理服务器

Notes

1. One of the most common methods of providing information assurance is to have an off-site backup of the data in case one of the mentioned issues arise. 一种最普通的提供信息保障

的方法是如果前面提到的问题发生时，有一个异地的数据备份！

2. Unfortunately, passwords are not foolproof, and hacking programs can run through millions of possible codes in just seconds. 不幸的是，口令不是万无一失的，而且黑客程序在几秒钟内可以浏览数以百万计的可能代码。

   not foolproof：不是万无一失的；run through：浏览；millions of：成千上万的。

3. FTP：File Transfer Protocol，文件传输协议，该协议使得主机间可以共享文件。FTP 使用 TCP 生成一个虚拟连接用于控制信息，然后再生成一个单独的 TCP 连接用于数据传输。

4. NDA：Non-Disclosure Agreement，这是买卖双方开始谈生意时的第一步。根据保护对象的不同分两种：单向 NDA（one-way NDA）和双向 NDA（two-way NDA）。单向保密协定只保护采购方的利益。

Part Ⅲ Activities

1. *Review Questions.*

Answer the following questions with information from the passage.

1）What aspects does information guarantee include?

2）How to set up password?

3）What is the function of the firewall?

4）What is the malware?

2. *Information to Complete.*

Complete the following sentences with the words and phrases from the passage.

1）A computer is any _____ with a processor and some memory. Such devices can range from non-networked _____ devices as simple as calculators, to networked mobile computing devices such as smartphones and _____.

2）The availability of _____, more powerful and less _____ computing equipment made electronic _____ processing within the reach of small business and the home user.

3）In essence, firewalls _____ the information that _____ through them, only allowing authorized content in.

3. Sentences to Translate.

Translate the following sentences into English.

1）值得注意的是，计算机并不一定意味着家里的台式电脑。

2）信息安全是保护可用性、隐私和数据的完整性的过程。

3）在19世纪中叶开发的更复杂的分类系统，允许政府根据敏感度管理他们的信息。

4）使用密码是提高信息安全的最基本的方法之一。

5）这些代码块被安装在电脑上窃取信息，限制了可用性，记录用户的行为，或破坏数据。

4. Matching and Description.

The passage you have read is about computers. The following are some useful expressions and explanations for computers. Read them carefully and find the items equivalent to those given in English in the table below. Then write the corresponding letter in the brackets.

1）a collection of information that is organized so that it can easily be accessed, managed, and updated

2）mathematical calculations and algorithmic schemes that transform plaintext into cyphertext

3）someone who seeks and exploits weaknesses in a computer system or computer network

4）any software used to disrupt computer operation, gather sensitive information, or gain access to private computer systems

5）a self-replicating virus that does not alter files but resides in active memory and duplicates itself

6）a software or hardware-based network security system that controls the incoming and outgoing network traffic based on applied rule set

a. 蠕虫　　　　　　　　　　　　（　）（　）hacker
b. 数据加密　　　　　　　　　　（　）（　）data encryption
c. 防火墙　　　　　　　　　　　（　）（　）database
d. 黑客　　　　　　　　　　　　（　）（　）worm
e. 数据库　　　　　　　　　　　（　）（　）firewall
f. 恶意软件　　　　　　　　　　（　）（　）malware

Part IV Further Reading

Advanced Authentication—CA Technologies

Concern about identity theft, data breaches and fraud is increasing but at the same time organizations are feeling pressure to enable employees, partners and customers to access more sensitive information from anywhere and any device. These market dynamics make multi-factor authentication and fraud prevention critical parts of any organization's security strategy.

CA Advanced Authentication is a flexible and scalable solution that incorporates both risk-based authentication methods like device identification, geolocation and user activity, as well as, a wide variety of multi-factor, strong authentication credentials. This solution can allow the organization to create the appropriate authentication process for each application or transaction. It can be delivered as on-premise software or as a cloud service and it can protect application access from a wide range of endpoints including all of the popular mobile devices. This comprehensive solution can enable your organization to cost effectively and enforce the appropriate method of strong authentication across environments without burdening end users.

- CA Auth Minder is a versatile authentication server that allows you to deploy and enforce a wide range of strong authentication methods in an efficient and centralized manner. It enables secure online interaction with your employees, customers and citizens by delivering multi-factor strong authentication for both internal and cloud-based applications. It includes mobile authentication applications and SDKs as well as several forms of out-of-band authentication.

- CA Risk Minder offers your organization multi-factor authentication that can detect and block fraud in real-time, without any interaction with the user. It integrates with any online application, including websites/portals and VPNs and analyzes the risk of online access attempts and transactions. This transparent form of multi-factor authentication utilizes contextual factors such as Device ID, geolocation, IP address and user activity information to calculate a risk score and recommend the appropriate action.

- CA Cloud Minder Advanced Authentication is a versatile authentication service that includes multifactor credentials and risk evaluation to help avoid inappropriate access and fraud. It can help you easily deploy and manage a variety of authentication methods to protect your users without the traditional implementation, infrastructure and maintenance costs.

• CA Payment Security is comprised of payment industry security products. Advanced authentication techniques facilitate the safe use of payment vehicles including Card Not Present (CNP), 3-D Secure, EMV CAP/DPA, mobile wallet and mobile OTP transactions. CA Payment Security enables organizations to improve security, reduce payment fraud, and increase customer adoption and acceptance—all while reducing friction in payment transactions.

What you have learned in this unit:

What you have learned in this unit:

Unit 8
Multimedia and Computer Games

After reading this unit and completing the exercises, you will be able to

☆ understand major characteristics of multimedia;

☆ be familiar with platform characteristics of computer games;

☆ identify contemporary gaming;

☆ apply your knowledge when using your computers.

Part I Lead-in

1. *Look at the pictures and have a discussion about them. Then produce the possible words and expressions that are related to the theme.*

Picture 1

Picture 2

Picture 3

Picture 4

Write down the relevant words and expressions in the space provided below.

_____

_____

2. ***Listen to the passage and then fill in the blanks with the exact words or phrases you have just heard.***

  Alongside the Internet, computer _____ are a perennial member of this axis of _____ evil. In some ways understandably so; they can _____ be incomprehensibly violent, and kids seem to be way too into them. Given this, it's not _____ surprising that somebody has decided they might be addictive. Not in the regular sense of something you want "just one more" of, _____ Pringles or episodes of 24, but in the proper, clinical, drink/drugs/gambling sense.

Part Ⅱ Read and Understand

1. ***Discuss the questions with your partners.***

☆ What does multimedia contain?

☆ How do we divide categories of multimedia?

☆ What is contemporary gaming?

☆ What are characteristics of computer game?

2. ***Time to read it.***

↳ Multimedia ↲

  Multimedia refers to content that uses a combination of different content forms. This contrasts with media that use only rudimentary computer displays such as text-only or traditional forms of printed or hand-produced materials. Multimedia includes a combination of text, audio, still images, animation, video, or interactivity content forms.

| Text | Audio | Still Images |
| Animation | Video | Interactivity |

Multimedia is usually recorded and played, displayed, or accessed by information content processing devices, such as computerized and electronic devices, but can also be part of a live performance. Multimedia devices are electronic media devices used to store and experience multimedia content. Multimedia is distinguished from mixed media in fine art; by including audio, for example, it has a broader scope.

Categorization of Multimedia

Multimedia may be broadly divided into linear and non-linear categories. Linear active content progresses often without any navigational control for the viewer such as a cinema presentation. Non-linear uses interactivity to control progress as with a video game or self-paced computer-based training.

Multimedia presentations can be live or recorded. A recorded presentation may allow interactivity via a navigation system. A live multimedia presentation may allow interactivity via an interaction with the presenter or performer.

Major Characteristics of Multimedia

Multimedia presentations may be viewed by person on stage, projected, transmitted, or played locally with a media player. A broadcast may be a live or recorded multimedia presentation. Broadcasts and recordings can be either analog or digital electronic media technology. Digital online multimedia may be downloaded or streamed. Streaming multimedia may be live or on-demand.

Multimedia games and simulations may be used in a physical environment with special effects, with multiple users in an online network, or locally with an offline computer, game system, or simulator.

The various formats of technological or digital multimedia may be intended to enhance the users' experience, for example to make it easier and faster to convey information. Or in entertainment or art, to transcend everyday experience.

A lasershow is a live multimedia performance.

Enhanced levels of interactivity are made possible by combining multiple forms of media content. Online multimedia is increasingly becoming object-oriented and data-driven, enabling applications with collaborative end-user innovation and personalization on multiple forms of content over time. Examples of these range from multiple forms of content on Web sites like photo galleries with both images (pictures) and title (text) user-updated, to simulations whose co-efficients, events, illustrations, animations or videos are modifiable, allowing the multimedia "experience" to be altered without reprogramming. In addition to seeing and hearing, haptic technology enables virtual objects to be felt. Emerging technology involving illusions of taste and smell may also enhance the multimedia experience.

Computer Games

PC games, also known as computer games, are video games played on a general-purpose personal computer rather than a dedicated video game console or arcade machine. Their defining characteristics include a lack of any centralized controlling authority and greater capacity in input, processing, and output.

PC games reached widespread popularity following the video game crash of 1983, particularly in Europe, leading to the era of the "bedroom coder". From the mid-1990s onward they lost mass-market traction to console games before enjoying a resurgence in the mid-2000s through digital distribution.

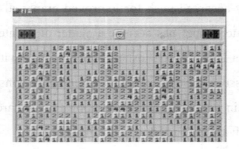

Microsoft Minesweeper

Contemporary Gaming

By 1996, the rise of Microsoft Windows and success of 3D console titles such as Super Mario 64 sparked great interest in hardware accelerated 3D graphics on the IBM PC compatible, and soon resulted in attempts to produce affordable solutions with the ATI Rage, Matrox Mystique and S3 ViRGE. Tomb Raider, which was released in 1996, was one of the first 3D third-person shooter games and was praised for its revolutionary graphics. As 3D graphics libraries such as DirectX and OpenGL matured and knocked proprietary interfaces out of the market, these platforms gained greater acceptance in the market, particularly with their demonstrated benefits in games such as Unreal. However, major changes to the Microsoft Windows operating system, by then the market leader, made many older MS-DOS-based games unplayable on Windows NT, and later, Windows XP (without using an emulator, such as DOS box).

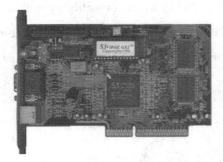 

S3 ViRGE GX2 graphics card   Tomb Raider

The faster graphics accelerators and improving CPU technology resulted in increasing levels of realism in computer games. During this time, the improvements introduced with products such as ATI's Radeon R300 and NVidia's GeForce 6 Series have allowed developers to increase the complexity of modern game engines. PC gaming currently tends strongly toward improvements in 3D graphics.

Unlike the generally accepted push for improved graphical performance, the use of physics engines in computer games has become a matter of debate since announcement and 2005 release of the nVidia PhysX PPU, ostensibly competing with middleware such as the Havok physics engine. Issues such as difficulty in ensuring consistent experiences for all players, and the uncertain benefit of first generation PhysX cards in games such as Tom Clancy's Ghost Recon Advanced Warfighter and City of Villains, prompted arguments over the value of such technology.

Platform Characteristics

Fidelity

In high-end PC gaming, a PC will generally have far more processing resources at its disposal than other gaming systems. Game developers can use this to improve the visual fidelity of their game relative to other platforms, but even (and in fact particularly) if they do not, games running on PC are likely to benefit from higher screen resolution, higher framerate, and anti-aliasing. Increased draw distance is also common in open world games.

Better hardware also increases the potential fidelity of a PC game's rules and simulation. PC games often support more players or NPCs than equivalents on other platforms and game designs which depend on the simulation of large numbers of tokens (e.g. Total War, Spore, Dwarf Fortress) are rarely seen anywhere else.

The PC also supports greater input fidelity thanks to its compatibility with a wide array of peripherals. The most common forms of input are the mouse/keyboard combination and gamepads, though touchscreens and motion controllers are also available. The mouse in particular lends players of first-person shooter and real-time strategy games on PC great speed and accuracy.

Gamepad

Openness

The defining characteristic of the PC platform is the absence of centralized control; all other gaming platforms (except Android devices, to an extent) are owned and administered by a single group.

The advantages of openness include:

- Reduced software cost

Prices are kept down by competition and the absence of platform-holder fees. Games and services are cheaper at every level, and many are free.

- Increased flexibility

PC games decades old can be played on modern systems, through emulation software if need be. Conversely, newer games can often be run on older systems by reducing the games' fidelity and/or scale.

- Increased innovation

One does not need to ask for permission to release or update a PC game or to modify an existing one, and the platform's hardware and software are constantly evolving. These factors make PC the center of both hardware and software innovation. By comparison, closed platforms tend to remain much the same throughout their lifespan. But there are also disadvantages, including:

- Increased complexity

A PC is a general-purpose tool. Its inner workings are exposed to the owner, and misconfiguration can create enormous problems. Hardware compatibility issues are also possible.

- Increased hardware cost

PC components are generally sold individually for profit (even if one buys a pre-built machine), whereas the hardware of closed platforms is mass-produced as a single unit and often sold at a loss. Games will also use PC hardware less efficiently than that of a closed platform, since the huge variety of possible makes and models prohibits focused optimisation.

- Reduced security

It is difficult, and in most situations ultimately impossible, to control the way in which PC hardware and software is used. This leads to far more software piracy and cheating than closed platforms suffer from.

Words & Expressions

analog /ˈænəlɒg/ adj. 模拟的
streaming multimedia 流媒体
simulation /ˌsɪmjuˈleɪʃn/ n. 仿真；模拟
enhance /ɪnˈhɑːns/ v. 提高；增强
convey /kənˈveɪ/ v. 传送；输送
illustration /ˌɪləˈstreɪʃn/ n. 插图；图解
animation /ˌænɪˈmeɪʃn/ n. 动画制作
modifiable /ˈmɒdɪfaɪəbl/ adj. 可修饰的
haptic /ˈhæptɪk/ adj. 触觉的
haptic technology 触感技术
emerge /iˈmɜːdʒ/ v. 浮现；摆脱；暴露
arcade machine 街机
console game 主机游戏
hardware accelerated 硬件加速
third-person shooter game 第三人称射击游戏
game engine 游戏引擎
middleware /ˈmɪdlweə/ n. 中间件
fidelity /fɪˈdeləti/ n. 保真；逼真度
anti-aliasing 抗锯齿的

Notes

1. This contrasts with media that use only rudimentary computer displays such as text-only or traditional forms of printed or hand-produced material. 这与早期的计算机只显示纯文本或传统形式的打印以及手工材料的媒体相比有着鲜明的对比。
   这里的 traditional forms of printed 是为了和 3D 打印相区别，特意表示为 traditional form，后面的 hand-produced material 指的是早期的计算机输出内容不是在显示器上，而是如纸带等手工材料上。
2. arcade machine：街机，是置于公共娱乐场所的经营性专用游戏机，也可称为大型电玩，起源于美国的酒吧。在街机上运行的游戏叫街机游戏。1971 年，世界第一台街机在美国的电脑试验室中诞生。在 20 世纪 80 年代的街机全是站立游戏的，进入 20 世纪

90年代出现控制面板略低的街机，配备座（椅）进行游戏。

3. console game：主机游戏，也就是中国玩家所说的"电视游戏"。英文中称游戏机为 video game console（视频游戏控制台），因为早期的游戏机类似与一个机器控制台，所以就用 console（控制台）指游戏机。console game 是指通过游戏机输出画面和声音到电视等音视频设备上的视频游戏。主机游戏机比较知名的有索尼的 PS 系列以及微软的 Xbox 系列。

4. ATI Rage，Matrox Mystique and S3 ViRGE：ATI Rage 是 ATI 公司早期推出的显卡，ATI 公司是世界著名的显示芯片生产商，在1985年至2006年是全球重要的显示芯片公司，总部设在加拿大安大略省，直至2006年被美国 AMD 公司以54亿美元的巨资收购后成为该公司的一部分。Matrox Mystique 和 S3 ViRGE 分别是 Matrox 公司和 S3 推出的显卡。

5. Tomb Raider：古墓丽影。1996年，英国 EIDOS 公司推出了一款名为《古墓丽影》（*Tomb Raider*）的游戏，由当时已经赢得了欧洲年度游戏奖的一个制作小组 Core Design 开发。当时的游戏首席图像设计师已经厌烦了硬汉加主视点射击风格的游戏，他提出创作一个女性冒险英雄的概念，再加上英国贵族的风格，劳拉（Lara）原型出现了。在当时3D技术还不太成熟的环境下，古墓丽影已经具备了一个成功游戏应该具备的一切，美式风格的女英雄，印第安纳琼斯的历险，加上在当时可谓是极为优秀的3D画面和操作，让这个游戏在发售了没多久就引起了全世界广大游戏玩家的浓厚兴趣。

6. NPC：Non-Player Character 的缩写，即非玩家控制角色。这个概念最早起源于单机版游戏，逐渐延伸到整个游戏领域。例如，在买卖物品的时候需要点击的那个商人就是 NPC，很多游戏里面需要对话的人物也属于 NPC。

7. gamepad：手柄。一种电子游戏机的输入设备，通过操纵其按钮等，实现对电脑上模拟角色等的控制。采用的就是家用游戏机式的手柄设计，左侧为方向键、右侧有4～6个功能键，根据需要还可能在别的部位加入更多的功能键，实现不同的功能。采用手柄比较适于进行模拟器类游戏，特别是一些滚屏类游戏。手柄在英文中有 gamepad 和 joystick 两个词，区别在于 gamepad 是手握的，joystick 是平放在桌上（和街机一样，主要用于街机模拟器和飞行类游戏）。

Part Ⅲ Activities

1. *Review Questions*.

Answer the following questions with information from the passage.

1）Do you know why multimedia is divided into linear and non-linear categories?

2) What are characteristics of multimedia?

3) When did PC games reach widespread popularity?

2. Information to Complete.

Complete the following sentences with the words and phrases from the passage.

1) Multimedia games and simulations may be _____ in a physical environment with special effects, with _____ users in an online network, or locally with an offline computer, game system, or simulator.

2) Unlike the _____ accepted push for improved graphical performance, the use of physics engines in computer games has become a matter of _____ since announcement and 2005 release of the nVidia PhysX PPU, ostensibly competing with middleware such as the Havok physics engine.

3) PC games decades old can be played on modern systems, through _____ if need be. Conversely, newer games can often be run on older systems by _____ _____ the games' fidelity and/or scale.

3. Sentences to Translate.

Translate the following sentences into English.

1) 多媒体可以大致分为线性和非线性类。

2) 更快的图形加速卡和 CPU 性能的提高最终提升了电脑游戏的真实感。

3) 多种媒体的结合使得提高交互的水平成为可能。

4) 多亏与外围设备的兼容性，PC 还支持更多的精确输入。

4. Matching and Description.

The passage you have read is about computers. The following are some useful expressions and explanations for computers. Read them carefully and find the items equivalent to those given in English in the table below. Then write the corresponding letter in the brackets.

1) the process of creating a continuous motion and shape change illusion by means of the rapid display of a sequence of static images that minimally differ from each other

2) a form of interactive multimedia used for entertainment that consists of manipulable images (and usually sounds) generated by a vide game console and displayed on a television or similar audio-video system

3) an artifact that depicts or records visual perception, for example a two-dimensional picture, that has a similar appearance to some subject—usually a physical object or a person, thus providing a depiction of it

4) a type of game controller held in two hands, where the fingers (especially thumbs) are used to provide input

5) a video game played over some form of computer network

a. 图片
b. 动画
c. 手柄
d. 主机游戏
e. 网络游戏

( ) ( ) animation
( ) ( ) image
( ) ( ) gamepad
( ) ( ) console game
( ) ( ) online game

Part IV Further Reading

History

The evolution of computer gaming can be traced back to tic-tac-toe simulators on 1940s cathode ray tube devices, as well as very basic titles played on mainframe machines. Gaming truly exploded into the public domain, though, with the advent of coin-op machines in arcades, which began to appear in the early 1970s. Arcade titles such as "Pong" by Atari and "Death Race" were popular. The first home gaming console, the Magnavox Odyssey, appeared in 1972. From there, both console and PC gaming began to escalate in terms of both popularity and power, with big names such as Microsoft, Sony, Nintendo and Sega all releasing home consoles. By 2011, PCs and gaming consoles support titles that feature vast worlds and highly advanced 3D graphics.

Game Formats

Video gaming can be divided roughly into two spheres: PC gaming, involving personal computers such as laptop devices, and console gaming. PC gaming utilizes a computer's mouse and keyboard to direct the action on-screen, and includes everything from flash-technology games played over the Internet to complex titles purchased from stores. Gaming consoles utilize video display signals, with the in-game action taking place on a TV screen.

Twenty-first-century consoles allow access to the Internet, a feature not found on earlier machines. Portable devices such as the Nintendo DS as well as cell phones, which contain a small screen and compact controls, allow for handheld gaming.

Genres

Among games, several genres have achieved popularity and continue to influence new titles. These include shooter games, which find players blasting enemies on the screen, and often use a first-person perspective, as seen in titles such as "Quake" and "Halo." Platform games, such as the "Super Mario" series, see players directing characters across 2D or 3D environments, with precise timing used to avoid enemies and obstacles. In role-playing games, such as "Fallout", players create a character and take part in quests, often developing their character as they progress. Sports titles simulate real-life activities, and include everything from Formula One to professional wrestling games.

Multiplayer

Many games allow more than one player to participate, typically simultaneously. In the 21st century, multiplayer gaming usually takes place via the Internet, with players all in their own home, with their own copies of the game. Popular Internet games include "World of Warcraft", a role-playing title enjoyed by thousands of players who interact online. Gaming consoles allow two or more players to participate at once, however, by using multiple control devices attached to one machine.

※ Unit 8   Multimedia and Computer Games ※

What you have learned in this unit:

What you have learned in this unit:

Unit 9
E-business

After reading this unit and completing the exercises, you will be able to

☆ understand major characteristics of e-business;

☆ be familiar with the security risks of e-business;

☆ identify what is the B2C;

☆ apply your knowledge when shopping online.

Part I Lead-in

1. *Look at the pictures and have a discussion about them. Then produce the possible words and expressions that are related to the theme.*

Picture 1

Picture 2

Picture 3

Write down the relevant words and expressions in the space provided below.

_____

_____

2. *Listen to the passage and then fill in the blanks with the exact words or phrases you have just heard.*

A final way to _____ information online would be to _____ _____ a digital signature. If a document has a digital signature on it, no one else is able to _____ the information without being detected. That way if it is edited, it may be adjusted for reliability after the fact. In order to use a digital signature, one must use a combination of cryptography and a _____ digest. A message digest is used to give the document a unique _____. That value is then encrypted with the sender's private key.

Part II Read and Understand

1. *Discuss the questions with your partners.*
☆ What's the e-business?
☆ Do you agree that online shopping is convenient and secure?
☆ How to ensure the security of electronic commerce?
☆ Why does a business enjoy a low overhead?

2. *Time to read it.*

↪ E-business ↩

E-business is the application of information and communication technologies in support of all the activities of business. Commerce constitutes the exchange of products and services between businesses, groups and individuals and can be seen as one of the essential activities of any business. Electronic commerce focuses on the use of ICT to enable the external activities and relationships of the business with individuals, groups and other businesses. The term "e-business" was coined by IBM's marketing and Internet teams in 1996.

E-business

Electronic Commerce

Basically, Electronic Commerce (EC) is the process of buying, transferring, or exchanging products, services, and/or information via computer networks, including the internet. EC can also be beneficial from many perspectives including business process, service, learning, collaborative, community. EC is often confused with e-business.

Classification by Provider and Consumer

Roughly dividing the world into providers/producers and consumers/clients, one can classify e-businesses into the following categories:

- Business-to-Business (B2B)
- Business-to-Consumer (B2C)
- Business-to-Employee (B2E)
- Business-to-Government (B2G)
- Government-to-Business (G2B)
- Government-to-Government (G2G)
- Government-to-Citizen (G2C)
- Consumer-to-Consumer (C2C)
- Consumer-to-Business (C2B)

Potential Concerns

While much has been written of the economic advantages of Internet-enabled commerce, there is also evidence that some aspects of the Internet such as maps and location-aware services may serve to reinforce economic inequality and the digital divide. Electronic commerce may be responsible for consolidation and the decline of mom-and-pop, brick and mortar businesses resulting in increases in income inequality.

Security

E-business systems naturally have greater security risks than traditional business systems; therefore it is important for e-business systems to be fully protected against these risks. A far greater number of people have access to e-businesses through the Internet than would have access to a traditional business. Customers, suppliers, employees, and numerous other people use any particular e-business system daily and expect their confidential information to stay secure. Hackers are one of the great threats to the security of e-businesses. Some common security concerns for e-businesses include keeping business and customer information private and confidential, authenticity of data, and data integrity. Some of the methods of protecting e-business security and keeping information secure include physical security measures as well as data storage, data transmission, antivirus software, firewalls, and encryption to list a few.

Privacy and Confidentiality

    Confidentiality is the extent to which businesses makes personal information available to other businesses and individuals. With any business, confidential information must remain secure and only be accessible to the intended recipient. However, this becomes even more difficult when dealing with e-businesses specifically. To keep such information secure means protecting any electronic records and files from unauthorized access, as well as ensuring safe transmission and data storage of such information. Tools such as encryption and firewalls manage this specific concern within e-business.

Security & Privacy

Data Integrity

    Data integrity answers the question "Can the information be changed or corrupted in any way?" This leads to the assurance that the message received is identical to the message sent. A business needs to be confident that data is not changed in transit, whether deliberately or by accident. To help with data integrity, firewalls protect stored data against unauthorized access, while simply backing up data allows recovery should the data or equipment be damaged.

Non-repudiation

    This concern deals with the existence of proof in a transaction. A business must have assurance that the receiving party or purchaser cannot deny that a transaction has occurred, and this means having sufficient evidence to prove the transaction. One way to address non-repudiation is using digital signatures. A digital signature not only ensures that a message or document has been electronically signed by the person, but since a digital signature can only be created by one person, it also ensures that this person cannot later deny that they provided their signature.

Access Control

    When certain electronic resources and information are limited to only a few authorized individuals, a business and its customers must have the assurance that no one else can access the systems or information. Fortunately, there are a variety of techniques to address this concern including firewalls, access privileges, user identification and authentication techniques (such as passwords and digital certificates), Virtual Private Networks (VPN), and much more.

Availability

This concern is specifically pertinent to a business' customers as certain information must be available when customers need it. Messages must be delivered in a reliable and timely fashion, and information must be stored and retrieved as required. Because availability of service is important for all e-business websites, steps must be taken to prevent disruption of service by events such as power outages and damage to physical infrastructure. Examples to address this include data backup, fire-suppression systems, Uninterrupted Power Supply (UPS) systems, virus protection, as well as making sure that there is sufficient capacity to handle the demands posed by heavy network traffic.

UPS Prolink PRO 903s

↪ B2C E-commerce ↩

B2C stands for business to consumer. It differs from B2B which is business to business. The business to consumer company model is one in which companies offer their products or services for sale to consumers rather than to other businesses, as in B2B. Today, B2C often refers to the online selling of products, which is known as e-tail. E-tailing products to consumers may be conducted by either manufacturers or retailers.

Virtually any product can be e-tailed. The challenge for retailers and manufactures in a B2C online selling environment is to get consumer traffic to their website marketplace. This need makes Search Engine Marketing (SEM) of extreme importance in business to consumer selling. SEM promotes a website by making the site highly visible on search engines.

Consumers typically only choose websites on the first few pages of results that come up after they enter a keyword or phrase query into a search engine. In order to get their website ranked on the first few pages, businesses typically purchase paid listings as well as employ Search Engine Optimization (SEO) techniques that include using popular consumer keywords in their web text. The more Internet search traffic that can reach an e-tail website, the better because the business is receiving views from consumers actually interested in their product, industry or subject.

Providing on-site security is a must for B2C. Consumers must feel secure in entering credit card and bank information when ordering and paying for items. In spite of its challenges, the business to consumer online model can greatly benefit both consumers and businesses. Consumers get the convenience of shopping online in their own home, while businesses enjoy a low overhead since a storefront and a large inventory aren't necessary. Business can offer good prices to consumers and provide free or low cost shipping in many cases.

Since no brick and mortar storefront or large overhead costs are required, the B2C Internet model of selling is ideal even for small businesses. For instance, traditionally, a small jewelry designer wanting to sell to consumers would have mostly craft fairs as a venue. Today, with business websites so affordable to create, small businesses can thrive on selling their products directly to consumers via the Internet. Whereas a brick and mortar jewelry store would need to be filled with products for customers to look at, online photographs allow them to select items. This is a big advantage to a small business owner such as a jewelry designer, because the product doesn't have to be assembled until the consumer orders it.

 ## Words & Expressions

commerce /'kɒmɜːs/ n. 贸易
essential /ɪ'senʃl/ n. 要素；实质
roughly /'rʌfli/ adv. 大约；大致
evidence /'evɪdəns/ n. 证据
aspect /'æspekt/ n. 方面；外观
reinforce /ˌriːɪn'fɔːs/ v. 加固；给……加强力量（或装备）
confidential /kɒnfɪ'denʃl/ adj. 机密的
encryption /ɪn'krɪpʃn/ n. 加密
confidentiality /ˌkɒnfɪˌdenʃɪ'æləti/ n. 机密
recipient /rɪ'sɪpiənt/ n. 接受者
e-tail /'iːteɪ/ n. 电子零售

Notes

1. A far greater number of people have access to e-businesses through the Internet than would have access to a traditional business. 比起传统的商业，更多的人通过互联网进行电子商务交易。

   这里是一个 greater… than… 结构。

2. Some common security concerns for e-businesses include keeping business and customer information private and confidential, authenticity of data, and data integrity. 一些常见的电子商务安全问题，包括保证业务和客户信息的保密，数据的真实性以及数据的完整性。

   keep... private and confidential 译为保密，data integrity 译为数据完整性。

3. With any business, confidential information must remain secure and only be accessible to the intended recipient. 任何的商业行为，机密信息必须保证安全，同时只能被指定的接收者访问。

   only be accessible to the intended recipient 的主语是 confidential information。

4. Business-to-Consumer（B2C）：企业和顾客（个人）之间的交易，如天猫。而 C2C 的买卖主体都是顾客，淘宝是最典型的 C2C 站点，京东则属于 B2C。

5. VPN：虚拟专用网，是一种常用于连接中、大型企业或团体与团体间的私人网络的通讯方法。虚拟专用网络的信息透过公用的网络架构（例如：互联网）来传送内联网的网络信息。它利用已加密的通道协议（Tunneling Protocol）来达到保密、传送端认证、信息准确性等私人信息安全效果。这种技术可以用不安全的网络（例如：互联网）来传送可靠、安全的信息。需要注意的是，加密信息与否是可以控制的。没有加密的虚拟专用网信息依然有被窃取的危险。

6. UPS：不间断电源。UPS 是在电网异常（如停电、欠压、干扰或浪涌）的情况下不间断地为电器负载设备提供后备交流电源，维持电器正常运作的设备。通常情况下不间断电源被用于维持计算机（尤其是服务器）或交换机等关键性商用设备或精密仪器的不间断运行，防止计算机数据丢失，电话通信网络中断或仪器失去控制。

7. SEM：SEM 是 Search Engine Marketing 的缩写，中文意思是搜索引擎营销，是一种新的网络营销形式。所做的就是全面而有效地利用搜索引擎来进行网络营销和推广。

Part III Activities

1. *Review Questions*.

Answer the following questions with information from the passage.

1）What are the classifications of e-business by provider and consumer?

2）How do we deal with the existence of proof in a transaction of e-business?

3）How to ensure electronic commerce information security?

2. *Information to Complete*.

Complete the following sentences with the words and phrases from the passage.

1）However, this becomes _____ more difficult when dealing with e-businesses specifically. To _____ such information secure means any electronic records and files from unauthorized access, as well as ensuring safe transmission and data storage of such information.

2）A business must have _____ that the receiving party or purchaser cannot deny that a transaction has occurred, and this means having sufficient evidence to _____ the transaction.

3）_____ typically only choose _____ on the first few pages of results that come up after they enter a keyword or phrase query into a _____ engine.

3. *Sentences to Translate*.

Translate the following sentences into English.

1）电子商务活动中，参与交易的双方、金融机构都应当维护电子商务的安全、通畅与便利。

2）在电子商务中，公司可以用电子形式将关键的商务处理过程连接起来，以形成虚拟企业。

3）零售商和制造商面临的挑战是在 B2C 在线销售环境下，如何让消费者来他们的网站。

4）由于没有实体店面以及大额支出的成本，B2C 模式的销售方式对于小企业是非常理想的。

4. Matching and Description.

The passage you have read is about computers. The following are some useful expressions and explanations for computers. Read them carefully and find the items equivalent to those given in English in the table below. Then write the corresponding letter in the brackets.

1) a Chinese website for online shopping similar to eBay and Amazon that is operated in China by Alibaba Group
2) the largest bank in the world by total assets and market capitalization; which is one of China's "Big Four" state-owned commercial banks
3) a computer network that uses Internet Protocol technology to share information, operational systems, or computing services within an organization
4) a widely used cryptographic hash function producing a 128-bit (16-byte) hash value, typically expressed in text format as a 32 digit hexadecimal number
5) a cryptographic network protocol for secure data communication, remote command-line login, remote command execution, and other secure network services between two networked computers

a. MD5
b. 安全外壳协议
c. 淘宝
d. 工商银行
e. 企业内部网

(　) (　) SSH
(　) (　) ICBC
(　) (　) encryption
(　) (　) TaoBao
(　) (　) Intranet

Part IV Further Reading

Physical Security of E-business

Despite e-business being business done online, there are still physical security measures that can be taken to protect the business as a whole. Even though business is done online, the building that houses the servers and computers must be protected and have limited access to employees and other persons. For example, this room should only allow authorized users to enter, and should ensure that "windows, dropped ceilings, large air ducts, and raised floors" do not allow easy access to unauthorized persons. Preferably these important items would be kept in an air-conditioned room without any windows.

Protecting against the environment is equally important in physical security as protecting against unauthorized users. The room may protect the equipment against flooding by keeping all equipment raised off of the floor. In addition, the room should contain a fire extinguisher in case of fire. The organization should have a fire plan in case this situation arises.

In addition to keeping the servers and computers safe, physical security of confidential information is important. This includes client information such as credit card numbers, checks, phone numbers, etc. It also includes any of the organization's private information. Locking physical and electronic copies of this data in a drawer or cabinet is one additional measure of security. Doors and windows leading into this area should also be securely locked. Only employees that need to use this information as part of their job should be given keys.

Important information can also be kept secure by keeping backups of files and updating them on a regular basis. It is best to keep these backups in a separate secure location in case there is a natural disaster or breach of security at the main location.

"Failover sites" can be built in case there is a problem with the main location. This site should be just like the main location in terms of hardware, software, and security features. This site can be used in case of fire or natural disaster at the original site. It is also important to test the "fail over site" to ensure it will actually work if the need arises.

State of the art security systems, such as the one used at Tidepoint's headquarters（巴尔的摩司令部）, might include access control, alarm systems, and closed-circuit television. One form of access control is face (or another feature) recognition systems. This allows only authorized personnel to enter, and also serves the purpose of convenience for employees who don't have to carry keys or cards. Cameras can also be placed throughout the building and at all points of entry. Alarm systems also serve as an added measure of protection against theft.

Unit 9 E-business

What you have learned in this unit:

What you have learned in this unit:

Unit 10
Future of Computer

After reading this unit and completing the exercises, you will be able to

☆ understand the meaning of Moore's law;

☆ be familiar with the size of different SD cards;

☆ apply your knowledge when using your computers.

Part I Lead-in

1. *Look at the pictures and have a discussion about them. Then produce the possible words and expressions that are related to the theme.*

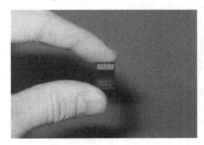

Picture 1

Picture 2

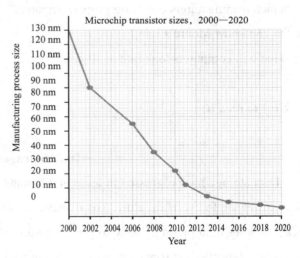

Picture 3

Write down the relevant words and expressions in the space provided below.

_____

_____

2. Listen to the passage and then fill in the blanks with the exact words or phrases you have just heard.

Today's computers operate using semiconductors, metals and electricity. Future computers might use atoms, dna or light. Moore's Law predicts doubling, but when computers go _____ quartz to quantum, the _____ will be off the scale.

What would the world be like, if computers the size of molecules become a _____ _____? These are the types of computers that could be everywhere, but never seen. Nano sized bio-computers that could target specific areas _____ your body. Giant networks of computers, in your clothing, your _____, your car. Entrenched in _____ every aspect of our lives and yet you may never give them a single thought.

Part II Read and Understand

1. *Discuss the questions with your partners.*

☆ How many times will micro-SD storage capacity be that of human brains, by 2030?

☆ Which manufacturers can design supercomputers?

☆ Do you know Moore's Law?

☆ What will computers look like in 30 years?

2. *Time to read it.*

↳**Data Storage**↵

Data storage has progressed in leaps and bounds over the last 50 years and will continue to do so for the foreseeable future. Trends have consistently shown exponential growth in this area, making it surprisingly easy to predict future advances. Before we look at these future advances, however, it is perhaps worth looking back at the history of data storage.

In 1956, IBM launched the RAMAC 305—the first computer with a Hard Disk Drive (HDD). This weighed over a ton and consisted of fifty 24" disks, stacked together in a wardrobe-sized machine. Two independent access arms moved up and down to select a disk, and in and out to select a recording track, all under servo control. The total storage capacity of the RAMAC 305 was 5 million 7-bit characters, or about 4.4 MB.

The IBM RAMAC 305

1962 saw the release of the IBM 1311—the first storage unit with removable disks. Each "disk pack" held around two million characters of information. Users were able to easily switch files for different applications.

Transistor technology—which replaced vacuum tubes—began to substantially reduce the size and cost of computer hardware.

The IBM 3330 was introduced in 1970, with removable disk packs that could hold 100 MB. The 1973 model featured disk packs that held 200 MB. Access time was 30 ms and data could be transferred at 800 KB/s.

Floppy disks arrived in 1971, revolutionizing data storage. Although smaller in capacity, they were extremely lightweight and portable. The earliest versions measured 8 inches in diameter. These were followed by 5¼-inch disks in the late 1970s and 3½-inch disks in the mid-1980s.

Floppy Disk

The IBM 3380 was introduced in 1980. This mainframe held eight individual drives, each with a capacity of 2.5 GB. The drives featured high-performance "cache" memory and transfer speeds of 3 MB/s. Each cabinet was about the size of a refrigerator and weighed 550 lb (250 kg). The price ranged from $648,000 to $1,136,600.

The growth of home computing in the 1980s led to smaller, cheaper, consumer-level disk drives. The first of these was only 5 MB in size. By the end of the decade, however, capacities of 100 MB were common.

Microcomputer

Data storage continued to make exponential progress into the 1990s and beyond. Floppy disks were replaced by CD-ROMs, which in turn were replaced by DVD-ROMs and began to be superseded by the Blu-ray format. Home PCs with 100 GB hard drives were common by 2005 and 1 terabyte (TB) hard drives were common by 2010.

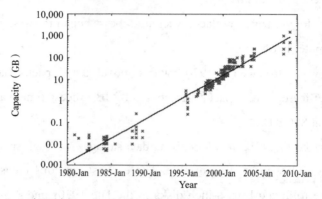

Secure Digital (SD) cards arrived in the early 2000s. They provided storage in a thumbnail-sized form factor, enabling them to be used with digital cameras, phones, MP3 players and other handheld devices.

Micro-SD cards (pictured below) have shrunk this format to an even smaller size. As of 2010, it is possible to store 32 GB of data on a device measuring 11 mm ×15 mm, weighing 0.5 grams and costing under $100.

To put this in context: this is over 3 million times lighter and over 10,000 times cheaper than an equivalent device of 30 years ago.

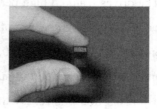

Micro-SD

So, what does the future hold?

It is safe to assume that the exponential trends in capacity and price performance will continue. These trends have been consistent for over half a century. Even if the limits of miniaturization are reached with current technology, formats will become available that lead to new paradigms and even higher densities. Carbon nanotubes, for example, would enable components to be arranged atom-by-atom.

The memory capacity of the human brain has been estimated at between one and ten terabytes, with a most likely value of 3 terabytes. Consumer hard drives are already available at this size.

128 GB micro-SD cards are being planned for 2011 and there is even a 2 TB specification in the pipeline.

Well before the end of this decade, it is likely that micro-SD cards (such as that pictured above) will exceed the storage capacity of the human brain.

By 2030, a micro-SD card (or equivalent device) will have the storage capacity of 20,000 human brains.

By 2043, a micro-SD card (or equivalent device) will have a storage capacity of more than 500 billion gigabytes—equal to the entire contents of the Internet in 2009.

By 2050—if trends continue—a device the size of a micro-SD card will have storage equivalent to three times the brain capacity of the entire human race.

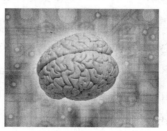

➥Internet Users➥

By 2020, the number of Internet users will reach almost 5 billion—equal to the entire world's population in 1987. This compares with 1.7 billion users in 2010 and only 360 million in 2000.

Vast numbers of people in developing countries will gain access to the web, thanks to a combination of plummeting costs and exponential improvements in technology. This will include laptops that can be bought for only a few tens of dollars, together with explosive growth in the use of mobile broadband. Even some of the most remote populations on Earth will gain access to the Internet.

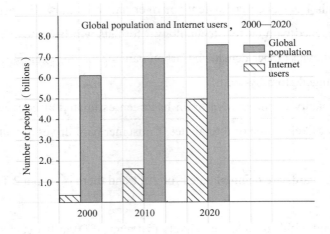

→Moore's Law←

Moore's Law describes a long-term trend in the history of computing hardware. The number of transistors that can be placed inexpensively on an integrated circuit doubles approximately every two years. This trend has continued in a smooth and predictable curve for over half a century and is expected to continue beyond 2020.

The capabilities of many electronic devices are strongly linked to Moore's Law: processing speed, memory capacity, sensors and even the pixels in digital cameras. All of these are improving at exponential rates as well. This is dramatically enhancing the impact of digital electronics in nearly every segment of the world economy.

In 2011, Intel unveiled a new microprocessor based on 22 nanometer process technology. Codenamed Ivy Bridge, this is the first high-volume chip to use 3D transistors, and packs almost 3 billion of them onto a single circuit. These new "Tri-Gate" transistors are a fundamental departure from the two-dimensional "planar" transistor structure that has been used before. They operate at much lower voltage and lower leakage, providing an unprecedented combination of improved performance and energy efficiency. Dramatic innovations across a range of electronics from computers to mobile phones, household appliances and medical devices will now be possible.

Even smaller and denser chips based on a 14 nm process are being planned for 2013, and the company's long-term roadmap includes sizes down to 4 nm in the early 2020s—close to the size of individual atoms. This will present major design and engineering challenges, since transistors at these dimensions will be substantially affected by quantum tunneling (a phenomenon where a particle tunnels through a barrier).

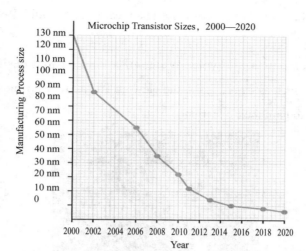

From the 2020s onwards, it is possible that carbon nanotubes or a similar technology will reach the mass market, creating a new paradigm that allows Moore's Law to continue. Chips constructed on an atom-by-atom basis would reach incredible densities.

Further into the future, chips may become integrated directly with the brain, combining AI/human intelligence and dramatically enhancing our cognitive and learning abilities. This could allow technologies once considered the stuff of science fiction to become a reality—such as full immersion VR, electronic telepathy and mind uploading.

Ultimately, Moore's Law could lead to a "technological singularity"—a point in time when machine intelligence is evolving so rapidly that humans are left far, far behind.

↪Supercomputers↩

Supercomputers are very large groups of computers that work together, combining their abilities to perform tasks that individual desktop computers would be incapable of. These include highly intensive calculations such as problems involving quantum mechanics, weather forecasting, climate research, astronomy, molecular modeling and physical simulations (such as simulation of aeroplanes in wind tunnels, simulation of nuclear weapons detonations and research into nuclear fusion).

Supercomputers were first developed in the 1960s. They were designed primarily by Seymour Cray at Control Data Corporation, which led the market into the 1970s, until Cray left to form his own company, Cray Research. He then took over the supercomputer market with his new designs, holding the top spot in supercomputing for five years (1985—1990).

In the 1980s, a number of smaller competitors entered the market, in parallel to the creation of the "mini-computer" market, but many disappeared in the mid-1990s supercomputer market crash.

Today, supercomputers are typically one-of-a-kind custom designs—produced by traditional companies such as Cray, IBM and Hewlett Packard, who had purchased many of the 1980s companies to gain their experience.

Roadrunner, a supercomputer built by IBM, in 2008.
It became the first 1.0 petaFLOP system.

Since October 2010, China has been home to the world's fastest supercomputer. The Tianhe-1A supercomputer, located at the National Supercomputing Center in Tianjin, is capable of 2.5 petaFLOPS; that is, over 2½ quadrillion (two and a half thousand million million) floating point operations per second.

America will take the lead once again in 2012, when IBM's Sequoia supercomputer comes online. This will have a maximum performance of 20 petaFLOPS; nearly an order of magnitude faster than Tianhe-1A. IBM believes it can achieve exaflop-scale computing by 2019, a thousandfold improvement over machines of 2010.

For decades, the growth of supercomputer power has followed a remarkably smooth and predictable trend, as seen in the graph below. If this exponential trend continues, it is likely that complete simulations of the human brain and all of its neurons will be possible by 2025. In the early 2030s, supercomputers could reach the zeta flop scale, meaning that weather forecasts will achieve 99% accuracy over a two-week period. By the 2050s, supercomputers may be capable of simulating millions—even billions—of human brains simultaneously. In parallel with developments in artificial intelligence and brain-computer interfaces, this could enable the creation of virtual worlds similar in style to the sci-fi movie, The Matrix. It is even possible that we are living in such a simulation at this very moment, without realizing it.

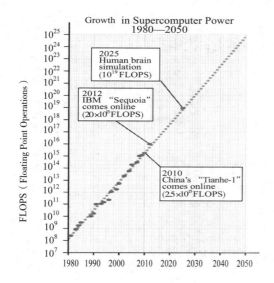

 Words & Expressions

leap /liːp/ n. 跳跃；剧增
foreseeable /fɔːˈsiːəbl/ adj. 可以预知的；能预见的
exponential /ˌekspəˈnenʃl/ adj. 指数的
predict /prɪˈdɪkt/ v. 预言；预报
servo /ˈsɜːvəʊ/ n. 伺服系统
revolutionize /ˌrevəˈluːʃənaɪz/ v. 彻底改变
supersede /ˌsuːpəˈsiːd/ v. 取代
terabyte /ˈterəbaɪt/ n. 万亿字节；太字节
handheld /ˈhændheld/ adj. 掌上型的
paradigm /ˈpærədaɪm/ n. 典范；样式
nanotube /ˈnænəʊtjuːb/ n. 纳米碳管
pipeline /ˈpaɪplaɪn/ n. 管道
exceed /ɪkˈsiːd/ v. 超过；超越
plummeting /ˈplʌmɪtɪŋ/ adj. 直线下降的
explosive /ɪkˈspləʊsɪv/ adj. 突增的；猛增的
inexpensively /ˌɪnɪkˈspensɪvli/ adv. 廉价地
dramatically /drəˈmætɪkli/ adv. 戏剧地；突然；巨大地
nanometer /ˈnænəʊmiːtə/ n. 纳米
high-volume 大容量

leakage /ˈli:kɪdʒ/ n. 泄漏
substantially /səbˈstænʃəli/ adv. 基本上；大体上
quantum /ˈkwɒntəm/ n. 量子
tunnel /ˈtʌnl/ v. 挖隧道
phenomenon /fəˈnɒmɪnən/ n. 现象
particle /ˈpɑ:tɪkl/ n. 颗粒
barrier /ˈbæriə/ n. 障碍物
cognitive /ˈkɒgnətɪv/ adj. 认知的
telepathy /təˈlepəθi/ n. 心灵感应
neuron /ˈnjʊərɒn/ n. 神经元
simultaneously /ˌsɪmlˈteɪniəsli/ adv. 同时地
sci-fi 科幻

Notes

1. Floppy disks were replaced by CD-ROMs, which in turn were replaced by DVD-ROMs, which in turn began to be superseded by the Blu-ray format. 软盘被 CD-ROM 替代，随之 CD-ROM 被 DVD-ROMs 取代，进而 DVD-ROMs 又被蓝光格式所取代。

   这里有连续两个定语从句，前面一个 which 修饰 CD-ROMs，后面一个 which 修饰 DVD-ROMs。

2. 128 GB micro-SD cards are being planned for 2011 and there is even a 2 TB specification in the pipeline. 128 G 的 micro-SD 卡计划在 2011 年上市，同时 2 TB 的规格也正在筹划中。

   这里的 planned for 省略了 selling，pipeline 本意为"管道"，这里扩展含义为"筹划，计划"。

3. Vast numbers of people in developing countries will gain access to the web, thanks to a combination of plummeting costs and exponential improvements in technology. 由于成本下降以及技术上极大的改进，广大发展中国家的人们得以访问互联网。

   "Vast numbers of people in developing countries will gain access to the web" 为主句。但是在翻译中，为了上下文的流畅，将句子的结构进行了调整。

4. Secure Digital（SD）cards：SD 卡。一种存储卡，被广泛地用于便携式设备，例如数码相机、个人数码助理和多媒体播放器等。SD 卡的技术基于 Multimedia 卡格式上。SD 卡有比较高的数据传送速度，而且不断更新标准。大部分 SD 卡的侧面设有写保护控制，以避免一些数据意外地写入，而少部分的 SD 卡甚至支持数字版权管理的技术。前几年很多手机的扩展卡都是使用的 SD 卡，后来随着 TF 卡（micro SD，体积更小）

的发展，现在的手机仅支持 TF 卡。SD 卡现在更多地出现在数码相机上。

5. Moore's Law：摩尔定律。该定律是由英特尔创始人之一戈登·摩尔提出来的，其内容为：集成电路上可容纳的电晶体数目，约每隔 24 个月便会增加一倍；经常被引用的"18 个月"是由英特尔首席执行官 David House 所说：预计 18 个月会将芯片的性能提高一倍（即更多的晶体管使其更快）。尽管这种趋势已经持续了超过半个世纪，摩尔定律仍应该被认为是观测或推测，而不是一个物理或自然法。

6. Tianhe-1A：天河 1A。全称天河一号超级计算机系统，是一台由中国人民解放军国防科学技术大学和天津滨海新区提供的异构超级计算机，天河的意思为银河。2010 年 10 月，经升级后的天河一号二期系统（天河-1A）以峰值速度（Rpeak）每秒 4 700 万亿次浮点运算、持续速度（Rmax）2 566 万亿次，超越橡树岭国家实验室的"美洲虎超级计算机"（Rpeak：2 331 万亿次；Rmax：1 759 万亿次），为当时世界上最快的超级计算机。2011 年 6 月，日本超级计算机"京"以比天河一号快 3 倍、以每秒 8 162 万亿次的运算速度，取得世界上最快的超级计算机的宝座，天河居第二。

Part Ⅲ   Activities

1. *Review Questions.*

   Answer the following questions with information from the passage.

   1）How many bytes is a micro-SD card's storage capacity by 2030?
   2）Which electronic devices are strongly linked to Moore's law?
   3）Why do we say, it is likely that complete simulations of the human brain and all of its neurons will be possible by 2025?

2. *Information to Complete.*

   Complete the following sentences with the words and phrases from the passage.

   1）By 2050—if trends continue—a device the size of a ＿＿＿＿＿＿＿＿＿＿ will have storage equivalent to three times the brain ＿＿＿＿＿＿＿＿ of the entire human race.

   2）From the 2020s onwards, it is possible that carbon nanotubes or a similar technology will reach the mass market, creating a new paradigm that allows ＿＿＿＿＿＿＿＿ to continue.

   3）In 2011, Intel unveiled a new ＿＿＿＿＿＿＿＿＿＿ based on 22 nanometer process technology. Codenamed Ivy Bridge, this is the first high-volume chip to use 3D transistors, and packs almost 3 billion of them onto a single ＿＿＿＿＿＿＿＿.

3. Sentences to Translate.

Translate the following sentences into English.

1) Micro-SD 卡将 SD 格式带到了更小的尺寸上。

2) 伴随着移动宽带用户的爆炸性增长，膝上型电脑将会用几十美元就能够买到。

3) 摩尔定律描述了计算机硬件发展的历史以及趋势。

4) 最终，摩尔定律可能会导致技术奇点——一个时间点，机器智能的发展如此迅速，以至于人类离灭亡也不远了。

4. Matching and Description.

The passage you have read is about computers. The following are some useful expressions and explanations for computers. Read them carefully and find the items equivalent to those given in English in the table below. Then write the corresponding letter in the brackets.

1) It is the observation that, over the history of computing hardware, the number of transistors on integrated circuits doubles approximately every two years. The law is named after Gordon E. Moore, co-founder of Intel Corporation, who described the trend in his 1965 paper.

2) It is a supercomputer capable of an Rmax (maximum range) of 2.566 petaFLOPS. Located at the National Supercomputing Center in Tianjin, China, it was the fastest computer in the world from October 2010 to June 2011 and is one of the few Petascale supercomputers in the world.

3) It is a non-volatile memory card format for use in portable devices, such as mobile phones, digital cameras, GPS navigation devices, and tablet computers.

4) It is a computer at the frontline of contemporary processing capacity—particularly speed of calculation which can happen at speeds of nanoseconds.

5) It is a data storage device using integrated circuit assemblies as memory to store data persistently.

a. 固态硬盘
b. 超级计算机
c. 摩尔定律
d. 安全数字卡
e. 地球模拟器

（　）（　）supercomputer
（　）（　）secure digital card
（　）（　）Moore's Law
（　）（　）SSD
（　）（　）Earth simulator

Part IV Further Reading

Future Computer

Future computer technology trends suggest that mankind has not even begun to tap into computational power and networking. In a few years, cloud computing will be so old school. The PC will be what the mainframe is today. And, Web 5.0 will have finally kicked in for future computers.

Future Computer

Not too many years down the road people will drive their future car inside their future city and then go back to their future home and tell their future robots what needs to be done. In the near future, wireless technology and nanotechnology will have taken over the world.

Nano neural networks will be the standard. Our brains will be connected directly to this vast future computer technology that has been miniaturized, broadened and made to be more powerful. Sound like science fiction?

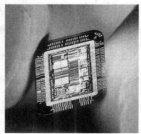

Right now in medical settings several quadriplegics have been fitted with microelectrodes in the motor cortex of their brains.

Just by thinking, they can move robotic arms. Microelectrodes placed in different parts of the brain allow people with muscular dystrophy and other disorders to move cursors and type keys rather adeptly on a computer.

No longer are these experiments done with rats, monkeys or the episode of "House" where a patient move a cursor for "yes" or "no" by thinking about it. There is so much future computer technology that most people don't know about that is happening right now, right under our noses.

And, so what does this bode for 5, 10, or 20 years from now? Trends in future computer technology dictate that microchips, processors, hard drives and all computer components will get smaller.

Wi-Fi, Mi-Fi and other Fi's Foes and Fums will take over the marketplace and become common. Just by thinking we will be able to connect with networks most likely aided by small electronic devices that we keep on us at all times.

But a step beyond this will be the elimination of these small electronic cell phone, palm pilot, laptop, DS, GPS, facial recognition devices that have been rolled into one. Networks will be so widespread that there will be no need to carry small handheld devices.

Just by thinking about logons, passwords, text, photos, music and whatever our heads hold we'll be able to navigate, communicate socially from miles away, tweet just by thinking and do work on the run.

Future computer technology will allow people who have lost the use of their limbs to use robotics as well as tap into these same networks empowering disabled people like never before.

Some level of Cyborg technology will develop (that will be both morally positive and morally questionable).

Future computer technology will help solve many medical problems such as dyslexia and ADHD by reinterpreting sensory data and modulating brain activity. A whole host of other medical illnesses and conditions will be cured or managed by future software controlling dosages of medicine such as those with diabetes now experience, but the scope of the conditions healed or managed will be much more vast.

Future computer software will know when to stimulate neurons and communicate with nano-robots that perform microscopic surgery or deliver medicine to very small, specific areas of the body.

The future of computer technology is a very bright one indeed. The current trends tell us this, as well as the research and development that is happening at lightning pace. The children today will have a whole new technological world waiting for them tomorrow. And tomorrow isn't that far away.

※ Unit 10　Future of Computer ※

What you have learned in this unit:

What you have learned in this unit:

Appendix A   Specialist English and Translation Skills

一、计算机英语与翻译技巧概述

翻译技巧在英语的学习中是非常重要的，翻译技巧就好像是架起两种语言的桥梁，桥梁的好坏直接影响了我们对两种语言的理解程度。在专业英语的翻译方面，这一点就尤其重要。

计算机在存储和显示信息时，通常尽量采用缩略的形式，使其语句十分简洁，加之专业术语较多，逐渐形成了人们常说的"计算机英语"。如果计算机英语像一般英语那样结构松散，书写完整，就会显得冗长啰唆，编写程序或阅读起来就会感到十分困难和诸多不便。尽管有的人学过多年英语，对绝大多数计算机英语句子能翻译、能理解，而对有些计算机英语词句就不知其所以然了。在计算机英语中有的语句按照字面意义、按照语法规则或习惯表达法去分析均视为错误，但在计算机英语中则是正确的表达方法。例如：

① Data record too large（数据记录太大），正常英语应写成：The data record is too large.
② Drive A not prepared（驱动器A没有准备好），正常英语应为：Drive A is not prepared.
③ Program too big to fit in memory（程序太大，不能装入内存），普通说法是：The Program is too big to fit in the memory.
④ General failure reading（writing）drive A（读写驱动器A失败），一般英语应写成：There is a general failure in reading（writing）drive A.
⑤ Write protect error writing drive A（写信息时驱动器A中软盘处于写保护状态），一般英语表达成：Drive A is in the status of writing protect when you are writing data on the floppy disk.

上面几个句子如按照字面意义或语法规则去分析，都或多或少有错，但在计算机英语中这些表达方式却是对的。我们不难发现，学好英语特别是学好计算机英语，不但可以加快掌握计算机知识，做到"既知其然又知其所以然"，而且还可以大大降低学习计算机的难度。所以在学习好基础英语的同时，非常有针对性地进行相关方面的练习，这样既可以学习好计算机英语，同时又可以巩固我们的通用英语，使我们的知识在两个方面都会有所提高。

二、专业英语翻译

随着现代科学技术的飞速发展，全球经济一体化逐步深入，科技英语将越来越引起科学界和语言界的高度重视和关注。著名科学家钱三强早就指出："科技英语在许多国家已经成为现代英语的一个专门的新领域。随着国际学术交流的日益广泛，专业英语已经受到普遍的重视，掌握一些专业英语的翻译技巧是非常必要的。科技英语（English for Science and Technology，简称 EST）作为一种重要的英语文体，与非专业英语文体相比，具有词义多、长句多、被动句多、词性转换多、非谓语动词多、专业性强等特点，这些特点都是由科技文献的内容所决定的。因此，专业英语的翻译也有别于其他英语文体的翻译。

（一）了解专业领域知识

在专业英语翻译中，要达到融会贯通，必须了解相关的专业，熟练掌握同一事物的中英文表达方式。单纯靠对语言的把握也能传达双方的语言信息，但运用语言的灵活性特别是选词的准确性会受到很大限制。要解决这个问题，翻译人员就要积极主动地熟悉这个专业领域的相关翻译知识。因此，了解了生产工艺和设备，在翻译过程中对语言的理解能力和翻译质量就会大大提高。

（二）专业英语翻译原则

关于翻译的原则，历来提法很多。有的主张"信、达、雅"，有的主张"信、顺"，有的主张"等值"等，并曾多次展开过广泛的争论和探讨。但是，从他们的争论中可以看出，有一点是共同的，即一切译文都应包括原文思想内容和译文语言形式这两个方面。简单地说，符合规范的译文语言，确切忠实地表达原作的风格，这就是英语翻译的共同标准。为此，在进行英语翻译时要坚持两条基本原则：

1. 忠实原文

译文应忠实于原文，准确地、完整地、科学地表达原文的内容，包括思想、精神与风格。译者不得任意对原文内容加以歪曲、增删、遗漏和篡改。

2. 文理通顺

译文语言必须通顺，符合规范，用词造句应符合本民族语言的习惯，要用民族的、科学的、大众的语言，以求通顺易懂。不应有文理不通、逐词死译和生硬晦涩等现象。

（三）专业英语翻译过程中要体现语言特色

1. 大量使用名词化结构

《当代英语语法》（*A Grammar of Contemporary*）在论述专业英语时提出，大量使用名词化结构（Nominalization）是专业英语的特点之一。因为科技文体要求行文简洁、表达客观、内容确切、信息量大、强调存在的事实，而非某一行为。例如：Computer software is a general term used to describe a collection of computer programs, procedures and documentation that perform some tasks on a computer system, including system software。（计算机软件是一个描述包括计算机程序，以及在计算机系统上执行特定任务的进程和记录如系统软件等集合的通用术语。）其中 collection 和 documentation 就是用的名词化结构。

2. 广泛使用被动语句

根据英国利兹大学 John Swales 的统计，专业英语中的谓语至少 1/3 是被动语态。这是因为科技文章侧重叙事推理，强调客观准确。第一、二人称使用过多，会造成主观臆断的印象。因此尽量使用第三人称叙述，采用被动语态，例如：The size of a computer monitor is measured diagonally, from the bottom left corner of the monitor to the top right corner. (计算机显示器的尺寸大小是从显示器的底部左角到顶部右角进行对角测量的。)

3. 非限定动词的应用和大量使用后置定语

如前所述，科技文章要求行文简练，结构紧凑，往往使用分词短语代替定语从句或状语从句；使用分词独立结构代替状语从句或并列分句；使用不定式短语代替各种从句；介词＋动名词短语代替定语从句或状语从句。这样可缩短句子，又比较醒目。

4. 大量使用常用句型

科技文章中经常使用若干特定的句型，从而形成科技文体区别于其他文体的标志。例如 It—that—结构句型、被动语态结构句型、分词短语结构句型、省略句结构句型等。举例如下：It is important that you clean your computer and your computer components periodically.（很重要的是你应该保持计算机和其外设组件的清洁。）

5. 为了描述事物精确，要使用长句

为了表述一个复杂概念，使之逻辑严密，结构紧凑，科技文章中往往出现许多长句。有的长句多达七八十个词。

6. 大量使用复合词与缩略词

大量使用复合词与缩略词是科技文章的特点之一，复合词从过去的双词组合发展到多词组合；缩略词趋向于任意构词，例如：hardware（硬件）、LCD（Liquid Crystal Display）（液晶显示器）、full-enclosed（全封闭的）。

（四）专业英语的翻译方法

要提高翻译质量，使译文达到"准确""通顺""简练"这三个标准，就必须运用翻译技巧。翻译技巧就是在翻译过程中用词造句的处理方法，如词义的引申、增减、词类转换和科技术语的翻译方法等。

三、英语词典

普通英语学习几年以后，虽然掌握了数千个单词的词汇量，但在专业领域的文章中，却有大量的专业词汇和术语晦涩难懂，这些词在文章中起着举足轻重的作用。光凭上下文含义猜测是行不通的。一本好的英语词典可以成为你的良师益友，在英语学习中起着不可忽视的作用。一本好的英语词典可以帮助学习者学习语言、澄清概念、养成良好的学习习惯。不少学习者都有过背词典的经历。因此正确的使用词典、善于使用词典，是英语学习的一个重要的环节。学习英语，查词典是必下的基本功夫。

虽说词典是最好的老师，但是英语学习者的具体情况千差万别，对于各式各样的词典，如果选择不当，这个"最好的老师"可能不仅不会有多少帮助，甚至可能会带来不少消极影响。因此正确选择适当的词典是英语学习的首要环节。

今天，市面上可供我们选择的词典可谓种类繁多，琳琅满目。有英文原版词典，也有各种英汉词典，以及英汉技术词典，专供学生学习的学习词典，还有英汉双解词典、图解词典、考试词典、语法词典、同义词词典等。面对如此丰富的选择，我们的确应该好好了解考察一番。

（一）英汉双解词典

英汉双解词典是在保留英语原版词典内容的基础上，为原版词典中的词条、例句和注释等提供对应的汉语翻译的词典。简单地说，它是在英语原版词典的基础上加上汉语释义的词典。比如《牛津英汉双解小词典》《牛津袖珍英汉双解词典》以及《牛津现代英汉双解词典》等都属于这类词典。

由于英语和汉语并不是完全对应的，因此，使用英汉双解词典，读者既能了解某单词在汉语中的对应词，又能通过其英语释义更清楚、更准确地理解其含义，避免单纯看汉语对应词而产生的词义扩大或缩小等带来的理解偏差。另外，使用英汉双解词典可以增加英语语感，同时学习到比较地道规范的英语用法和说法。

在查阅英汉双解词典时，不能只是满足于知道了一个英文单词的中文对应词，还要注意读它的英文原文的解释，以便更进一步领会其准确含义。

（二）英语学习词典

英语学习词典是专为母语不是英语的学习者编纂的英语词典，其特点是：
①选词适当，根据语料库数据分析词频后选用英语学习者最需要的常用词汇。
②释义简明，释义词汇控制在一定数量之内（如2 000～3 500个常用词汇），使学习者使用起来非常方便，因此很受英语学习者的欢迎。
③例句丰富，最新的英语学习词典都根据语料库的语料，提供生活中的真实例句。
④语法详解，还提供句型，疑难之处有重点说明。
⑤用法详细，提供使用说明、词汇搭配、词语辨析，学习功能齐全。

市面上常见的《朗文当代英语学习辞典》《朗文高阶英语词典》《朗文中阶英语词典》《麦克米伦高阶英语词典》《麦克米伦高阶美语词典》《外研社·建宏英汉多功能词典》《韦氏中国学生英汉双解词典》等都是很好的学习词典。

（三）英英词典

英英词典一般是由母语为英语的人编写的词典，它是纯英语的单语词典，无论是释义还是例句，都是用英语表达的，没有中文对应词和翻译。

英英词典最大的好处是提供了一个纯英语的环境。通过查词典可以增加英语的阅读量，而看懂英语释义的过程就是学习用英语思考、用英语理解的过程。如果要真正学好英语，至少要有一本英英词典。另外，使用英英词典能够增加语感，准确理解英语单词的词义及用法，有助于培养英语学习者书面语和口语的准确表达及语言技能的恰当运用。

英英词典是纯英语的单语词典，无中文释义，因此学习者需要英语水平较高，一般说来中级或中级以上水平的读者就可以使用英英词典。中级水平的学习者可以使用《朗文中阶英语词典》《剑桥英语学习词典》（英语版）；对高水平的学习者来说，《朗文当代英语辞典》（第三版增补本）、《朗文高阶英语词典》《麦克米伦高阶英语词典》和《麦克米伦高阶美语词典》等，都是很好的英英词典。

（四）汉英双语词典

汉英双语词典这个概念的提出，源于外语教学与研究出版社的《现代汉语词典》（汉英双语版）。这部双语词典是在《现代汉语词典》的基础上全文翻译而成，实际上相当于汉语词典的双解版。

（五）图解词典

图解词典就是以图解的形式来编排词典，主要词条配有图片，帮助查阅者准确理解词义，因此非常形象直观。图解字典可以加强对单词的感性认识，帮助记忆，又能增强查阅词典的趣味性，不知不觉学会了单词，且扩大了知识面。图解词典的适用对象可以是成人，也可以是儿童，但表现形式和内容有很大不同。《外研社·DK·牛津英语图解大词典》就是这种图解词典的典范。该词典图文并茂，为查阅者提供了科学技术、文化和自然历史等方面的全面信息，而且查阅方便。

（六）电子词典

电子词典以其独特的功能，如操作简单、携带方便、价格低廉等特点赢得了广大英语学习者的青睐，成为其工作学习的好帮手。

最初的电子词典类似于计算器，只不过增加了一些英汉词汇和记事功能。现在的电子词典有的收录英汉、汉英词汇高达几万、十几万，甚至二三十万；英语、汉语的词汇和语句达到真人发音效果；可以用键盘输入或手写输入；可以与计算机通讯，甚至上互联网发送电子邮件；可以采取插卡的方式查阅百科全书和各类专业词典；可以记名片，学英语，玩游戏……总之是品种繁多，功能各异，轻薄厚重，异彩纷呈。一台电子词典在手，可以代替许多办公学习用具，且携带方便，查询迅速。

一般来说非英语专业的大学生需要一本解释详尽的英汉双解词典、一本难度适中的英英词典和一本中型汉英词典。可选择的对象包括《牛津袖珍英汉双解词典》、《朗文当代高级英语辞典》（英英英汉双解）、《朗文当代高级英语辞典》（第三版增补本）、《麦克米伦高阶英语词典》（英语版）及《汉英词典》（修订版缩印本）或《现代汉英词典》（新版）。

英语学习者在选择词典时主要应注意以下几点：一看出版社是否知名，在专业领域是否有专长，这是质量的保证。二是看词典的出版目的，是供学习使用，还是仅供阅读时参考。词典一般都有封底文字的宣传，以及使用说明和体例说明，在购买时可以结合正文对比这些说明，看看是否名副其实。最好仔细阅读一些词条，来感受一下是否符合自己的需要。然后看词典的规模和收词量，决定它是不是适合自己的英语水平，一般选择略高于自己现有水平的词典。接着看词典的出版日期，尽量买新出版的词典，这样收词会比较新，如果有好几个版本，当然要买新版的了。最好要看看词典的装帧和印刷质量。因为词典需要经常翻阅使用，所以装帧的质量尤其重要，有的词典虽然价格便宜，但用不久就会破损，影响使用的心情和效果。优秀出版社的词典不光注意质量，在装帧和印刷方面也会精益求精。

在专业英语的学习过程中，还要正确使用技术词典或标准术语词典。技术词典的收词具有很强的时效性，多有专业侧重。但是，许多技术词典中的译名缺乏权威性，时常有不规范的说法，这在专业英语的学习及运用中是正常的现象。因此，学习者在使用词典的同时应该多关注一下最新的技术文献和标准化文件。特别应该注意是伴随着计算机科学技术的快速发展，计算机领域的专业术语时常都在更新和变化。例如IPX，IP，TCP/IP，IPV6，SPMT，POP，WEB，WEB 2.0，AJAX等。

四、英语翻译技巧（一）

英语翻译包括口译和笔译。在笔译中，又可分为科技翻译、文学翻译、政论文翻译和应用文翻译等。翻译就是"把一种语言文字的意义用另一种语言文字表达出来"（引自《现代汉语词典》）。因而，翻译本身并不是一门独立的创造性科学，它既带有创造性，又带有科学性，它是用语言表达的一门艺术，是科学性的再创作。

英语理解困难，这是学习和使用英语的人共同的感觉，这是由于东西方历史、文化、风俗习惯的不同，所以一句英语在英美人士看来顺理成章，而在我们看来却是颠颠倒倒、断断续续，极为别扭，很难理解。此外中文表达也很难，英译汉有时为了要找到一个合适的对等词汇，往往被弄得头昏眼花，想破头皮。还有翻译时要求尽量了解和掌握各种文化知识和背景，对比、概括和总结英汉两种不同语言的特点，以找出一般的表达规律来，避免出现一些不该出现的翻译错误，而这些表达的规律就是所说的翻译技巧。

（一）词义的选择和引申

英汉两种语言都有一词多类和一词多义的现象。一词多类就是指一个词往往属于几个词类，具有几个不同的意义；一词多义就是同一个词在同一词类中又往往有几个不同的词义。在英译汉的过程中，我们在弄清原句结构后，就要善于运用选择和确定原句中关键词词义的技巧，以使所译语句自然流畅，完全符合汉语习惯的说法；选择确定词义通常可以从两方面着手：

①根据词在句中的词类来选择和确定词义。

They are as like as two peas. 他们相似极了。（形容词）

He likes mathematics more than physics. 他喜欢数学甚于喜欢物理。（动词）

Wheat，oat，and the like are cereals. 小麦、燕麦等皆系谷类。（名词）

②根据上下文联系以及词在句中的搭配关系来选择和确定词义。

He is the last man to come. 他是最后来的。

He is the last person for such a job. 他最不配干这个工作。

He should be the last man to blame. 怎么也不该怪他。

This is the last place where I expected to meet you. 我怎么也没料到会在这个地方见到你。

词义引申是英译汉翻译时常用的技巧之一。翻译时，有时会遇到某些词在英语辞典上找不到适当的词义，如果任意硬套或逐词死译，就会使译文生硬晦涩，不能确切表达原意，甚至会造成误解。这时就应根据上下文和逻辑关系，从该词的根本含义出发，进一步加以引申，引申时，往往可以从三个方面来加以考虑。

①词义转译。当我们遇到一些无法直译或不宜直译的词或词组时，应根据上下文和逻辑关系，引申转译。

The energy of the sun comes to the earth mainly as light and heat. 太阳能主要以光和热的形式传到地球。

②词义具体化。根据汉语的表达习惯，把原文中某些词义较笼统的词引申为词义较具体的词。

The last stage went higher and took the Apollo into orbit round the earth. 最后一级火箭升得更高，把"阿波罗号"送进围绕地球运行的轨道。

③词义抽象化。根据汉语的表达习惯，把原文中某些词义较具体的词引申为词义较抽象的词，或把词义较形象的词引申为词义较一般的词。

Every life has its roses and thorns. 每个人的生活都有甜有苦。

（二）词性转译

在英译汉过程中，有些句子可以逐词对译，有些句子则由于英汉两种语言的表达方式不同，就不能逐词对译，只能将词类进行转译之后，方可使译文显得通顺、自然；对词性转译技巧的运用须从以下四个方面加以注意。

①转译成动词。英语中的某些名词、介词、副词，翻译时可转译成汉语中的动词。

The lack of any special excretory system is explained in a similar way.

植物没有专门的排泄系统，可用同样的方式加以说明。（名词转译）

As he ran out, he forgot to have his shoes on. 他跑出去时，忘记了穿鞋子。（副词转译）

②转译成名词。英语中的某些动词、形容词，翻译时可转换成汉语中的名词。

The earth on which we live is shaped like a ball.

我们居住的地球，形状像一个大球。（动词转译）

The doctor did his best to cure the sick and the wounded.

医生尽了最大的努力来治疗病号和伤员。（形容词转换）

③转译成形容词。英语中有些作表语或宾语的抽象名词，以及某些形容词派生的名词，往往可转译成汉语中的形容词。另外，当英语动词转译成汉语名词时，原来修饰该动词的副词也往往随之转译成汉语中的形容词。

It is no use employing radar to detect objects in water.

使用雷达探测水下目标是没有用的。（作表语的名词转译）

The sun affects tremendously both the mind and body of a man.

太阳对人的身体和精神都有极大的影响。（副词转译）

④转译成副词。英语中的某些名词、形容词，翻译时可转译成汉语中的副词。

When he catches a glimpse of a potential antagonist, his instinct is to win him over with charm and humor. 只要一发现有可能反对他的人，他就本能地要用他的魅力和风趣将这些人争取过来。（名词转译）

五、英语翻译技巧（二）

（一）增词翻译技巧

增词译法指英译汉翻译时，按意义上、修辞上或句法上的需要加一些词，使译文更加忠实通顺地表达原文的思想内容；但是，增加的并不是无中生有，而是要增加原文中虽无其词却有其意的一些词汇，这是英译汉翻译中常用的技巧之一。增词技巧一般分作两种情况。

①根据意义上或修辞上的需要，可增加表示名词复数的词、动词、副词、表达时态的词、语气助词和表达概括的词等六类词汇来补充表达。例如：

Flowers bloom all over the yard. 朵朵鲜花满院盛开。（增加表示名词复数的词）

After the banquets, the concerts and the table tennis exhibitor, he went home tiredly.

在参加宴会、出席音乐会、观看乒乓球表演之后，他疲倦地回到了家里。（增加动词）

He sank down with his face in his hands. 他两手蒙着脸，一屁股坐了下去。（增加副词）

He had known two great social systems.

那是以前，他就经历过两大社会制度。（增加表达时态的词）

As for me, I didn't agree from the very beginning.

我呢，从一开始就不赞成。（增加语气助词）

The article summed up the new achievements made in electronic computers, artificial satellites and rockets.

本文总结了电子计算机、人造卫星和火箭这三方面的新成就。（增加概括词）

②根据句法上的需要增补一些辅助性词汇。例如：

Reading makes a full man; conference a ready man; writing an exact man.

读书使人充实；讨论使人机智；写作使人准确。（增补原文句子中所省略的动词）

All bodies on the earth are known to possess weight.

大家都知道地球上的一切物质都有重量。（增补被动句中泛指性的词）

（二）正反、反正翻译技巧

正反、反正翻译是指翻译时突破原文的形式，采用变换语气的办法处理词句，把肯定的译成否定的，把否定的译成肯定的。运用这种技巧可以使译文更加合乎汉语规范或修辞要求，且不失原意。这种技巧可分五个方面加以陈述。

①肯定译否定。

The above facts insist on the following conclusions. 上述事实使人们不能不得出以下结论。

②否定译肯定。

She won't go away until you promise to help her. 她要等你答应帮助以后才肯走。

③双否定译肯定。

There can be no sunshine without shadow. 有阳光就有阴影。

但是，如果翻译时保留英语原来的"否定之否定"的形式并不影响中文的流畅时，则应保留，这样还可突出原文中婉转的语气。如 He is not unequal to the duty.（他并非不称职。）

④正反移位。

I don't think he will come. 我认为他不会来了。

⑤译为部分否定。

Not all minerals come from mines. 并非所有矿物都来自矿山。

Both of the substances do not dissolve in water. 不是两种物质都溶于水。

（三）翻译的重复技巧

重复技巧是英译汉翻译中的一种必不可少的翻译技巧。由于英译汉翻译时往往需要重复原文中的某些词才能使译文表达明确具体；又由于英汉语言结构不同，重复的手段和作用也往往不尽相同，大致可分为三种。

①为了明确表达意思。

I had experienced oxygen and /or engine trouble.

我曾碰到过，不是氧气设备出了故障，就是引擎出故障，或两者都出故障。（重复名词）

Under ordinary conditions of pressure, water becomes ice at 0 ℃ and steam at 100 ℃.

在常压下，水在零摄氏度时变成冰，在一百摄氏度时变成蒸汽。（重复动词）

A locality has its own over-all interest, a nation has another and the earth gets another.

一个地方有一个地方的全局，一个国家有一个国家的全局，一个地球有一个地球的全局。（重复谓语部分）

②为了强调语气。

He wandered along the street, thinking and thinking, brooding and brooding.

他在街头游来荡去，想了又想，盘算了又盘算。

③为了表达更加生动。

While stars and nebulae look like specks or small patches of light, they are really enormous bodies. 星星和星云看起来只是斑点，或者是小片的光，但它们确实是巨大的天体。

六、网上在线翻译

长期以来我们一般都在使用传统的纸质中英文词典或者电子词典来查询单词，这样不但需要经常把词典带在身边，而且在查询翻译的时候也会花费很多的时间。现在随着计算机技术迅速发展和网络技术的广泛普及，出现了很多可以在计算机系统上安装使用的英文翻译软件，例如常用的金山词霸、金山快译等翻译软件。不仅如此现在还可以实现免安装，使用浏览器在网上轻松进行在线翻译。只要我们的计算机系统连上互联网，就可以通过互联网在在线翻译平台来查找英文单词、词汇乃至进行整句整段的翻译。这种方式不仅方便、功能强大、收录词汇量巨大、释义丰富，还提供了多种类型的查询功能，如词义、例句、词组、同义词、反义词查询等，而且几乎都是免费的。例如，全球最大的互联网搜索引擎Google公司就提供了功能强大的在线多国语言互译和整句互译平台。

以下列举一些常用的、权威的、功能强大的网上在线翻译词典，供英语学习者参考。

英汉类词典
在线金山词霸　爱词霸
　　中国自主开发的最权威的免费在线词典，可查词、可在线翻译等，包含 70 多个专业的近百本词典数据，超过 1 000 万词条。
百度词典搜索
　　全球最大的中文搜索引擎百度推出的百度词典搜索支持强大的英汉互译功能，中文成语的智能翻译。
Google 语言工具
　　全球最大的搜索引擎 Google 公司推出的在线多国语言互译平台，功能强大。
Google 翻译
　　Google 公司推出的，目前网上最好的在线翻译工具。
Dict.cn 在线翻译字典
　　海词在线翻译搜索平台，支持词典查询、在线翻译、英语课堂、生词本、背单词等功能，可搜索不重复汉英词条 100 万，英汉词条 103 万。
Yahoo! 字典
　　全球知名互联网公司雅虎，提供了功能强大的在线中英文翻译平台，而且还推出了雅虎鱼等综合翻译工具，并针对不同地区推出差异性的在线翻译平台，足可见该公司对中英文市场的重视。
牛津英汉双解词典
　　全球最权威的英语词典出版社英国牛津大学出版社推出的在线英汉双解词典，界面简单而实用，释义准确简明，收词广泛经典。

缩写类词典
常用缩写
世界各国和地区名称代码表（标准代码，国际域名缩写）
行业代码表
Acronym Finder

百科全书类词典
C 书百科
　　在线词汇自由收录，现有 50 多万词汇。
大英百科全书
　　（英）世界著名的百科全书。
百度百科
　　这是一部由全体网民共同撰写的百科全书。
维基百科
　　维基百科，自由的百科全书。

计算机类词典

IT 用语词典

国内最专业、收词最丰富的计算机专业词汇在线查询平台，IT 用语的读法、意思、关联用语等可以用关键词搜索，类别目录以及拼音的 3 种方式来查找。

英汉计算机及网络通信技术词汇

这是一个个人网站，收录了一些计算机及网络通信技术专业词汇，并在不断地更新中。

CNPedia 华文电脑百科

由一家台湾的高科技企业开发推广的计算机专业词汇大全，是网络版的"计算机科学百科全书"。

趋势科技 - 病毒百科

Encyberpedia Dictionary and Glossary

（英文）因特网上主要的字典和词汇一览表。

On-line Dictionary of Computing

（英文）在线计算机专业词典。

Computer User High-Tech Dictionary

（英文）计算机用户使用的在线计算机专业词典。

Whatis_com, the leading IT encyclopedia and learning center

（英文）领先的 IT 技术百科全书和学习中心。

科技类词典

郑州大学在线英汉 - 汉英科技大词典

本词典固定词库中现收录英汉科技词语 750 269 条，汉英科技词语 744 542 条。

词博科技英语词典

收录 1 000 万科技词汇，搜索结果多元化。

包装词汇英汉对照

有关包装词汇的英汉对照词库，约 29 000 条。

英汉电子工程辞典。

Movie Terminology Glossary

电影专业术语辞典。

化工字典_中国试剂信息网

现有词汇量 351 681 条。其中包含 CAS 编号的有 246 435 条。

纺织字典

全球纺织网，收录有关纺织方面的词汇量高。

七、如何撰写电子邮件

电子邮件（electronic mail，简称 E-mail，标志: @，也被称为"伊妹儿"）又称电子信箱、电子邮政，它是一种用电子手段提供信息交换的通信方式，是因特网应用最广的服务：

通过网络的电子邮件系统，用户可以用非常低廉的价格（不管发送到哪里，都只需负担电话费和网费即可），以非常快速的方式（几秒钟之内可以发送到世界上任何你指定的目的地），与世界上任何一个角落的网络用户联系，这些电子邮件可以是文字、图像、声音等各种方式。同时，用户可以得到大量免费的新闻、专题邮件，并实现轻松的信息搜索。这是任何传统的方式也无法相比的。正是由于电子邮件的使用简易、投递迅速、收费低廉、易于保存、全球畅通无阻，使得电子邮件被广泛地应用，它使人们的交流方式得到了极大的改变。

随着我国经济的飞速发展，改革开放的不断深入，国际交流的日益增加，以及互联网技术的日新月异，学习如何撰写英文电子邮件对当今的人们来说就显得格外重要。现在简单介绍一下撰写英文电子邮件的方法和技巧。

首先从问候开始，用问候语开始邮件内容非常重要，例如"Dear Lillian,"。根据你与收件人的关系亲近与否，你可能选择使用他们的姓氏来称呼他们而不是直呼其名，例如"Dear Mrs. Price,"。如果关系比较亲密的话，你就可以说，"Hi Kelly,"。如果你和公司联系，而不是个人，你就可以写"To Whom It May Concern,"。

其次感谢收件人，如果你在回复客户的询问，你应该以感谢开头。例如，如果有客户想了解你的公司，你就可以说，"Thank you for contacting ABC Company."。如果此人已经回复过你的一封邮件了，那就一定要说，"Thank you for your prompt reply."或是"Thanks for getting back to me."。如果你可以找到任何机会那一定要谢谢收信人。这样对方就会感到比较舒服，而且显得更礼貌。

随后表明你的意图，如果是你主动写电子邮件给别人的话，那就不可能再写什么感谢的字句了，则以你写此邮件的目的开头。例如，"I am writing to enquire about..."或是"I am writing in reference to..."。在电子邮件开头澄清你的来意非常重要，这样才能更好地引出邮件的主要内容。记得要注意语法、拼写和标点符号，保持句子简短明了且句意前后一致。

在你结束邮件之前，再次感谢收信人并加上些礼貌语结尾。你可以"Thank you for your patience and cooperation."或"Thank you for your consideration."开始，接着写"If you have any questions or concerns, don't hesitate to let me know."及"I look forward to hearing from you."。

最后是写上合适的结尾并附上你的名字。"Best regards,""Sincerely,"及"Thank you,"都很规范化。最好不要用"Best wishes,"或"Cheers,"类的词，因为这些词都常用在非正式的私人邮件中。最后，在你发送邮件之前，最好再读一遍你的内容并检查其中有没有任何的拼写错误，这样就能保证你发出的是一封真正完美的邮件！

以下是一封来自一家美国知名的软件公司 EPICOR 向其合作伙伴发去的一封商业邮件，作为范例供参考。

From: "Siew Sun" <sschan@epicor.com>
Date: Mon, 12 Jan 2009 13:36:59 +0800（CST）
To: "Tina" < twang@epicor.com>
CC: "Angela Yap"<AYap@epicor.com>,"Anson"<alee@epicor.com>,"Vincent"<VTang@

epicor.com>,"Craig Charlton" <CCharlton@epicor.com>

Subject: New/Updated Education Courses

Hi all Vantage/Vista Partners,

Follow the link below to view the latest announcements and other communications related to Epicor products' education and documentation initiatives:

https://epicweb.epicor.com/Education/Pages/EdDocAnnounce.aspx

If you do not remember or know your EPICweb username and password, email issupportcenter@epicor.com.

Best regards,

Siew Sun

Channel Operations Administrator, APAC

Tel.: （603）7962-8830

Fax: （603）7722-4633

Cell: （6012）602-3005

E-Mail: sschan@epicor.com

Website: www.epicor.com

Partner Secured Site: https://epicweb.epicor.com/Partners

八、电脑游戏术语

电脑游戏（Personal Computer Games，Computer Games 或 PC Games）是指在电子计算机上运行的游戏软件。这种软件是一种具有娱乐功能的电脑软件。电脑游戏产业与电脑硬件、电脑软件、互联网的发展联系甚密。电脑游戏为游戏参与者提供了一个虚拟的空间，从一定程度上让人可以摆脱现实世界，在另一个世界中扮演真实世界中扮演不了的角色。同时计算机多媒体技术的发展，使游戏给了人们很多体验和享受。

电脑游戏的出现与 20 世纪 60 年代电子计算机进入美国大学校园有密切的联系。当时的环境培养出了一批编程高手。1962 年一位叫斯蒂夫·拉塞尔的大学生在美国 DEC 公司生产的 PDP-1 型电子计算机上编制的《宇宙战争》（*Space War*）是当时很有名的电脑游戏。一般认为，他是电脑游戏的发明人。

20 世纪 70 年代，随着电子计算机技术的发展，其成本越来越低。1971 年，被誉为"电子游戏之父"的诺兰·布什内尔发明了第一台商业化电子游戏机。不久他创办了世界上第一家电子游戏公司——雅达利公司（ATARI）。在 20 世纪 70 年代，随着苹果电脑的问世，电脑游戏才真正开始了商业化的道路。此时，电脑游戏的图形效果还非常简陋，但是游戏的类型化已经开始出现了。

从 20 世纪 80 年代开始，PC 机大行其道，多媒体技术也开始成熟，电脑游戏则成了这些技术进步的先行者。尤其是在 3Dfx 公司的 3D 显示卡给行业带来了一场图像革命以后，电脑游戏发展迅速。

进入 20 世纪 90 年代，电脑软硬件技术的进步，因特网的广泛使用为电脑游戏的发展带来了强大的动力。进入21世纪，网络游戏成了电脑游戏的一个新的发展方向。

因此，电脑游戏发展至今业已形成自己的一套专门的术语系统，现在特将一些常用游戏术语做一系统整理，列表以便学习者查阅。

3D Accelerator：3D加速器。一种专门提升PC的3D运算功能硬件，但其不能提升计算机整体的显示效果。

A

ACT（Action Game）：动作类游戏。这类游戏提供玩家一个训练手眼协调及反应力的环境及功能，通常要求玩家所控制的主角（人或物）根据周遭情况变化做出一定的动作，如移动、跳跃、攻击、躲避、防守等，来达到游戏所要求的目标。此类游戏讲究逼真的形体动作、火爆的打斗效果、良好的操作手感及复杂的攻击组合等。

AI（Artificial Intelligence）：人工智能，就是指计算机模仿真实世界的行为方式及人类思维与游戏的方式的运算能力。那是一整套极为复杂的运算系统与运算规则。

AVG（Adventure Game）：冒险类游戏。这类游戏在一固定的剧情或故事下，提供玩家一个可解谜的环境及场景，玩家必须随着故事的安排进行解谜。游戏的目的是借游戏主角在故事中所冒险积累的经验来解开制作者所设定的谜题或疑点。这类游戏常被用来设计成侦探类型的解谜游戏。

B

Boss：大头目，也称"老板"。在游戏中出现的较为巨大有力与难缠的敌方对手。一般这类敌人在整个游戏过程中只会出现一次，而常出现在关底，而不像小怪物在游戏中可以重复登场。

C

Cheat：游戏秘技。游戏设计者暗藏在作品中的特殊技巧，使用后可带给玩家特殊的能力与效果。最先是程序者为快速测试作品而设计的内部秘技，现在几乎是每个游戏均有秘技。

Clock Speed：游戏执行速度。游戏在计算机中被运行的速度，常以Megahertz（MHz）计量。

D

Doom-like：三维射击类游戏，即第一人称射击类游戏。游戏画面即为玩家的视野范围。现在此类游戏多称作Quake-like。

E

E3（Electronic Entertainment Expo）：美国E3大展。当前世界上最盛大的电脑游戏与电视游戏的商贸展示会，基本于每年五月举行。

Easter Egg：复活节彩蛋。程序中隐藏着的一段意外的内容，多为制作者设计的搞笑内容，经常是关于制作者自己的介绍与调侃。

ECTS（European Computer Trade Show）：欧洲计算机商贸展示会，被称为欧洲的E3大展，每年三月和九月于伦敦举行。

Electronic Game：电子游戏，即电脑游戏、电视游戏以及街机和手掌型游戏机的总称。

Engine：游戏引擎，即一套游戏的主程序。

※ Appendix A　Specialist English and Translation Skills ※

Experience Point：经验点数。常出现在角色扮演游戏中，以数值计量人物的成长，经验点数达到一定数值后常常会升级，这时人物就会变得更强大。

F

FTG（Fighting Game）：格斗类游戏。从动作类游戏脱胎分化出来的两个角色一对一决斗的游戏形式。现在此类游戏又分化出 2D 格斗类游戏与 3D 格斗类游戏。

First Person：第一人称视角，就是指屏幕上不直接出现主角，而是表现为主角的视野范围。

Flight Sim：飞行模拟类游戏。模拟类游戏下的一个门类，让玩家感受到操纵飞机以及飞翔于蓝天上的乐趣。

FMV（Full-Motion Video）：全动态影像，即游戏的片头、过场和片尾的动态画面。

FPS（Frames Per Second）：每秒显示帧数。美国的标准 NTSC 的电视节目的每秒显示帧数为 30。不少电脑游戏的显示帧数都超过了这个数字。

Free Guy：额外的命。在游戏中，你有可能会获得的额外的命，比如收集一定数量的某种宝物。

G

Game Over：游戏结束。这是游戏中最常见的话语，通常是表示游戏者失败，而不是通关爆机。

Gameplay：游戏可玩性，即游戏的玩法，是决定一个游戏有多好玩的重要因素。

Genre：游戏类型，即为不同游戏玩法的游戏作一归类，比如角色扮演类、冒险类、动作类、模拟类等。

Graphic Adventure：图形冒险类游戏。冒险类游戏下的一个门类，对应文字冒险类游戏。

H

Hidden Level：隐藏关卡。指游戏中隐藏的部分，需玩家自行发现。即不玩到这部分也能够通关，但玩到后可能会使情节起变化。

High-Res：高解析度，即精细的画面显示模式，但游戏的运行速度可能会因此有所下降。

Hint：攻略提示。简单的攻关提示，帮助玩家解决游戏中出现的特别棘手的难题。

HP（Hit Point）：生命力，即人物或作战单位的生命数值。一般 HP 为 0 即表示死亡，甚至 Game Over。

HUD（Heads Up Display）：飞行仪表盘。飞行模拟游戏中的常见词，常提供玩家诸如弹药状况、速度、目标跟踪等作战信息。

I

Interactive Movie：交互式电影，即结合游戏要素与电影要素的一类计算机互动作品，常常包含大量的 FMV（全动态影像）。

Interface：游戏界面，即玩家操作游戏的方式。它决定游戏的上手难度与可玩性。

J

Joypad：游戏手柄。模拟电视游戏的手柄，通常外接在声卡上。

Joystick：游戏操纵杆。常用来玩飞行、赛车等模拟类游戏的外接操纵杆。

L

Level：关卡，即游戏一个连续的完整的舞台、场景，有时也称作 Stage。

Low-Res：低解析度，即粗糙的画面显示模式，但运行速度可稍微提高。

M

Motion Capture：动态捕捉。将物体在 3D 环境中运动的过程数字化的过程。

Motion Tracker：动作跟踪器。动态捕捉时使用到的设备。

Moves：出招，即格斗游戏中人物的出招技巧。

MP（Magic Point）：魔法力，即人物的魔法数值，一旦使用完即不能再使用魔法招式。

MUD（Multi-User Dungeon）：多用户地牢，俗称"泥巴"。在互联网络上的一种允许多人参与的实时游戏，一般类似 RPG 的玩法，但目前多为文字模式。

N

Network Games：网络游戏，指容许多人通过某种网络协议连线后便能进行集体游戏的游戏种类。

NPC（Non Player Character）：非玩家人物。在角色扮演游戏中，玩家会在游戏过程中遭遇的所有不受控制的人物。这些人物或会提示重要情报线索，或是无关紧要的人物。

P

Password：过关密码。在游戏一开始处输入后便能直接进入后面的关卡。

Pirate：盗版游戏，即国内到处泛滥令游戏制作公司头痛不已的盗版游戏。

Platformer：游戏平台。游戏运行的平台，包括 Win95、DOS 或者 UNIX 等。

Player Killing：玩家杀手，指在 MUD 中，专以攻击玩家人物，而不是非玩家人物的一类玩家。

Polygon：多角形。运用在 2D 屏幕中表现 3D 环境的多角形单位。

Prototype：原型制作。游戏作品的原型制作也就是指以最快的速度制作出游戏的原型，一个可以执行的程序原型。从这些基础程序与基础图形，制作者可以看到从电脑中表现出来的与原来设想的有多大差距，经过调整磨合后就进入正式动工了。

Puzzle：谜题。在冒险类游戏中，考验玩家智力的谜题。

PZL（Puzzle Game）：解谜类游戏。一类专以不断解谜为主要内容的游戏种类。

Q

Quake-like：三维射击类游戏，即第一人称射击类游戏。见 Doom-like 条。

R

Round：回合。格斗类游戏中的一个较量的回合。

RPG（Role Playing Game）：角色扮演类游戏。这类游戏提供玩家一个可供冒险的世界（Fantasy World）或者一个反映真实的世界（Real World），这世界包含了各种角色、建筑、商店、迷宫及各种险峻的地形。玩家所扮演的主角便在这世界中通过旅行、交谈、交易、打斗、成长、探险及解谜来揭开一系列的故事情节线索，最终走向胜利的彼岸。玩家依靠自身的胆识、智慧和机敏获得一次又一次的成功，使自己扮演的主角不断发展壮大，从而得到巨大的精神满足。

RTS（Realtime Strategy Game）：即时战略类游戏。对应回合制战略游戏，一切都是实时发生，要求玩家具备较好的敏捷与宏观指挥能力。

S

Second Person：第二人称视角，即追尾视角，紧随游戏主角的背影。

Side-Scrolling：横向卷轴，即游戏画面的前景与背景从左向右移动的卷轴模式，常用于 2D 射击游戏中。

SLG（Simulation Game）：模拟类游戏。这类游戏提供玩家一个可以做逻辑思考及策略、战略运用的环境，且让玩家有自由支配、管理或统御游戏中的人、事或物的权力，并通过这种权力及谋略的运用达成游戏所要求的目标。玩家在条件真实、气氛宏大的游戏环境中充分施展智慧，克敌制胜，达到高层次的成功享受。

SPT（Sport）：运动类游戏。这类游戏提供一个反映现实（指正常的运动方式及运动精神）中的运动项目，并让玩家借助控制或管理游戏中的运动员或队伍，来进行运动项目的比赛。

Stage：关卡。见 Level 条。

STG（Shooting Game）：射击类游戏。有平面射击类与三维射击类（即第一人称射击类）。平面射击类还包括横向卷轴与纵向卷轴两种。射击类游戏是早期电脑游戏最常见的种类。

Storyline：剧情，即游戏的故事大纲，分为直线型、多线型以及开放型等三种。

Strategy Guide：战略指南手册，即游戏包装盒内附有的基本战略指导手册。

Sub-boss：隐藏头目。有些游戏中会隐藏有更厉害的大头目，通常是在通关后出现。

T

TAB（Table）：桌面类游戏。这类游戏提供一个训练逻辑思考或解谜的环境，并且有一定的规则及逻辑。玩家必须遵循游戏所设定的规则来解开谜题，达成游戏目标。此类游戏讲究高超的人工智能、新奇的玩法和舒适的操作环境。玩家在游戏中自得其乐、逍遥自在，也是一番享受。

TBS（Turn-Based Strategy Game）：回合制战略游戏。对应即时战略类游戏。参加战斗的几方，可以包括计算机在内，依一定顺序分别部署战略。一次部署便称作一个回合。

Tester：游戏测试者。游戏制作公司专门花钱聘请的测试作品的资深玩家。

Text Adventure：文字冒险类游戏。冒险类游戏下的一个门类，对应图形冒险类游戏，多是日本制作的小成本卡通游戏。

Third Person：第三人称视角。电脑游戏中最常见的视角，尤其是 2D 游戏。玩家是以第三者的角度观察场景与主角的动作。

V

Vertically Scrolling：垂直卷轴，即游戏画面的前景与背景从向下向上移动的卷轴模式，常用于 2D 射击游戏中。

VR（Virtual Reality）：虚拟实境。

W
Walkthrough：游戏攻略，是指完整的游戏攻关指导。
Z
Z-line/Z-axis：Z 轴。在 3D 环境中，Z 轴一般表示深度，X 轴表示高度，Y 轴表示宽度，而具备了 Z 轴就构成了 3D 环境。

九、网络聊天英语

互联网技术的日新月异的发展到今天，即时通信技术将人们之间交流和沟通的方式带向了新的境界。人们通过即时通信软件不论在何时何地，都能以所希望的方式，立即与所需联系的人即时沟通。它除了可发送文字信息之外，还可以拨打全球免费网络电话，在交谈的同时共享照片和文件，使用选择性隐身保护自己的隐私，用形象秀、表情符号和卡通娃娃尽情展现自我独有的风格。因此即时通信软件风靡全球，发展迅速，可以说目前会使用计算机进行工作的人没有说不用即时通信软件进行交流的。这也让全球各地的人们不分地域不分国界不分时间进行及时交流通讯了。

目前国际上经典的即时通信软件，包括 ICQ、MSN Messenger、雅虎通、AIM、Skype 等，而国内风靡流行的即时通信软件有腾讯 QQ、网易泡泡、新浪 668、搜 Q、朗玛 UC、IMU 即时通等。当人们大量使用即时通讯软件进行聊天沟通的时候，逐渐发展起来了特有的网络聊天语言，如果你不懂得这些特有的表达方式，那么就无法和英语国家的网友进行交流。现将一些目前流行的网络聊天英语罗列出来，方便英语学习者参考使用。

缩写	全称	中文
ASAP	as soon as possible	尽快
BF	boyfriend	男朋友
BTW	by the way	随便说一下
BBL	be back later	稍后回来
BRB	be right back	很快回来
CU	see you	再见
CUL	see you later	下次再会
DIIK	damned if I known	我真的不知道
DS	Dunce Smiley	笨伯
FE	for example	举例
FTF	face to face	面对面
FYI	for your information	供参考
GF	girlfriend	女朋友
IAE	in any event	无论如何
IC	I see	我明白
ILY	I love you	我爱你
IMHO	in my humble opinion	依愚人之见
IMO	in my opinion	依我所见
IOW	in other words	换句话说

LOL	laughing out loudly	大声笑
NRN	no reply necessary	不必回信
OIC	oh, I see	哦，我知道
PEM	privacy enhanced mail	保密邮件
RSVP	reply if you please	请答复
TIA	thanks in advance	十分感谢
TTUL	talk to you later	以后再讲
TY	thank you	谢谢
VG	very good	很好
WRT	with respect to	关于
WYMM	will you marry me	愿意嫁给我吗

Appendix B　Screen Prompt Message

在计算机系统的使用过程中屏幕上经常会出现一些使用英语书写的提示信息，大家很多都不知道是什么意思。这些提示信息是计算机人机交互操作中重要组成部分，一般是由操作系统或应用程序根据用户的操作而做出的反应。这些提示信息都是预先储存在系统内部的，为了降低储存量，它们都很短小精悍，在没有上下文和对系统具体运行机制不了解的情况下，是很难理解的。

本附录的列表中整理了一些常见的计算机系统出现问题时屏幕提示的错误信息。当大家遇到这些提示信息时，通过查阅下面的列表可以了解该提示信息的中文意思，并查看其解释说明，从而对症下药解决问题，排除故障。

屏幕提示信息	中文翻译	提示信息解释说明
Abort, Retry, Ignore, Fail?	退出，重试，忽略，取消？	不能识别给出的命令、或发生了使命令不能执行的磁盘或设备错误，可能是磁盘损坏或软驱门没关。按 A 键彻底终止，并回到 DOS 提示符；按 R 键重复执行该命令；按 I 键继续处理，忽略错误，非常冒险，建议不要采用；按 F 键不执行有问题的命令，继续下面的处理。
Access Denied	拒绝存取	试图打开一个标记为只读、存储在写保护的磁盘上或锁定在网络上的文件。如果在子目录上使用"Type"命令，或在文件上使用"CD (chdir)"命令，也会产生这个信息。应该用"Attrib"命令删除文件的只读状态或从磁盘中去掉写保护，然后再试试。
Auxiliary device failure	辅助设备出现故障	触摸板或鼠标可能出现故障。如果你仅使用鼠标，请检查连接处是否松动或有不正确连接的现象。如果问题仍然存在，请启用"Pointing Device"定点设备选项。
Bad command or file name	错误的命令或文件名，不能识别输入的命令	应该检查以确保输入命令的正确性，确认在指定目录或用 Path 命令指定的搜索路径上能找到命令文件。
Boot error	引导错误	在引导时检测不到应有的外设。应该检查计算机的设置参数，如用户自己不能解决这个问题，请找专门维修人员。

续表

屏幕提示信息	中文翻译	提示信息解释说明
Cache disabled due to failure	高速缓存出现故障被禁用	微处理器内部的主高速缓存出现故障。
Cannot find system files	不能找到系统文件	试图从没有包含系统文件的驱动器上装入操作系统。应该用"sys"命令将系统文件复制到根目录中。除非真的是不能恢复系统文件了，才可用 format/s 命令重新格式化磁盘。
Cannot load command, system halted	不能加载 command，系统中止应用程序覆盖了内存中的所有或部分 Command.com	应该重新引导计算机，检查被应用程序修改过的数据是否完整，如有必要可将 Command.com 复制到子目录，这样退出应用程序时 DOS 可在这儿找到 Command.com。
Cannot read file allocation table	不能读到文件分配表	文件分配表已坏。如仍能找到一些数据，那么将它们都备份到一张空盘中，也可利用 Chkdsk 命令修复文件分配表，如需要，可重新格式化磁盘。如果问题重复发生，那么应该修理驱动器或更换驱动器。
CD-ROM drive controller failure 1	CD-ROM 驱动器控制器故障 1	CD-ROM 驱动器无法对计算机发出的命令作出反应。关闭计算机，然后从介质托架连接器中断开 CD-ROM 驱动器的连接。重新启动计算机，然后再次关闭计算机，将 CD-ROM 驱动器重新连接至计算机，然后验证介质托架电缆是否连接至 CD-ROM 驱动器背面。重新启动计算机即可。
CMOS battery failed	CMOS 电池失效	一般出现这种情况就是说明给主板 CMOS 供电的电池已经快没电了，需要及时更换主板电池。
CMOS check sum error-defaults loaded	CMOS 执行全部检查时发现错误，要载入系统预设值	一般来说出现这句话有两种解释：一种是说主板 CMOS 供电电池快要没电了，可以先换个电池试试看；第二种解释是如果更换电池后问题还是没有解决，那么就说明 CMOS RAM 可能有问题了，如果主板没过一年的话就可以到经销商处换一块主板，要是过了一年就让经销商送回生产厂家修一下吧！

续表

屏幕提示信息	中文翻译	提示信息解释说明
Data error	数据错误	软盘或硬盘驱动器无法读取数据。运行适当的公用程序检查软盘驱动器或硬盘驱动器上的文件结构。
Decreasing available memory	可用内存正在减少	一个或多个内存模块可能出现故障或插接不正确。在升级插槽中重置内存模块。如果问题仍然存在，请从升级插槽中卸下内存模块。
Disk C: failed initialization	磁盘 C: 初始化失败	硬盘驱动器初始化失败。卸下并重置硬盘驱动器，然后重新启动计算机。如果问题仍然存在，请从诊断程序软盘中引导系统，然后运行 "Hard-Disk Drive"（硬盘驱动器）检测程序。
Diskette drive 0 seek failure	软盘驱动器 0 寻道失败	电缆可能已松动，或者系统配置信息可能与硬件配置不匹配。检查并重置软盘驱动器电缆。如果问题仍然存在，请运行 Dell 诊断程序中的 "Diskette Drive"（软盘驱动器）检测程序，并检查系统设置程序中相应驱动器的设置是否正确 ["Diskette Drive A"（A 软盘驱动器）或 "Diskette Drive B"（B 软盘驱动器）]。
Diskette read failure	读取软盘失败	电缆可能已松动，或者软盘可能出现故障。如果软盘驱动器访问指示灯可以亮起，请尝试使用其他软盘。
Diskette subsystem reset failed	软盘子系统重设失败	软盘驱动器控制器可能出现故障。
Diskette write-protected	软盘已写保护	由于软盘已启用写保护，操作可能无法完成。将写保护弹片向上滑动。
Divide overflow	分配溢出，除零错误	程序可能编写有错误，未调试好，也可能是与内存中的其他程序冲突。检查内存中的其他程序或不再使用此程序。
Drive not ready	驱动器未准备就绪	软盘驱动器中没有软盘，或者驱动器托架中没有硬盘驱动器。继续操作之前，请在软盘驱动器中装入软盘或在托架中安装硬盘驱动器。在软盘驱动器中装入软盘，或者将软盘推入软盘驱动器直至弹出按钮向外弹出，或者在驱动器托架中安装硬盘驱动器。
Drive not ready error	驱动器未准备好	没有该驱动器或未放磁盘。检查磁盘或更换磁盘。
Duplicate file name or file not found	文件重名或未找到	给文件起名字时与已有的文件重名了或是在对文件操纵时根本就没这条文件。更换名字或是检查文件名的拼写。

续表

屏幕提示信息	中文翻译	提示信息解释说明
Error loading operating system	引导操作系统错误	操作系统文件找不到或已损坏。用 SYS 命令将操作系统文件拷贝到该驱动器,如需要,可将 config.sys 和 autoexec.bat 文件拷贝到根目录中。如不能恢复系统文件,那么从软盘引导系统,备份数据,用 format/s 命令重新格式化磁盘。
Error reading PCMCIA card	读取 PCMCIA 卡错误	计算机无法识别 PC 卡。重新插接此卡,或者尝试使用另一张已知可以正常工作的 PC 卡。
EXEC failure	文件执行失败	应用程序的可执行文件包含影响处理的错误,或者由于早已打开了太多的文件而不能打开该文件,文件可能与当前的 DOS 版本不兼容。检查 DOS 的版本,如版本正确,可通过编辑 Config.sys 中的 Files 命令来解决这个问题。
Extended memory size has changed	已更改扩展内存大小	NVRAM 中记录的内存容量与计算机中安装的内存不符。重新启动计算机即可。
File allocation table bad	文件分配表已损坏	很多原因,例如病毒发作,突然停机,不正常关机等都能破坏分配表。将所能找到的数据备份到空盘中,不要覆盖以前的备份。也许可通过引用 Chkdsk 命令来解决这个问题。如需要,重新格式化软盘,如问题反复出现,那么将驱动器送去修理。
File cannot be copied onto itself	文件不能拷贝成自己	你在源文件和目标文件中指定了相同的文件,或是忘了写文件名。按需要改变源文件或目标文件,然后再试试看。
File creation error	文件建立错误	可能是在磁盘中没有足够的空间为用户创建文件、想创建的文件早已存在,且为只读文件或是想利用早已存在的文件名来更换文件的名字。可以换个盘,或使用别的目标名、别的目标位置,或者使用 Attrib 命令除去文件的只读属性。
File not found	文件未找到	在当前目录或由 Path、Append 命令指定的任一目录中找不到文件,或者指定的目录是空的。检查文件名的拼法和位置,如需要可以改变搜索路径。
Floppy Disk（s）fail 或 Floppy Disk（s）fail（80）或 Floppy Disk（s）fail（40）	无法驱动软盘驱动器	系统提示找不到软驱,首先要看看软驱的电源线和数据线有没有松动或者是接反,最好是把软驱放到另外一台机子上试一试,如果这些都不行,那么只好再买一个了,好在目前市场中的软驱还不算贵。

续表

屏幕提示信息	中文翻译	提示信息解释说明
Gate A20 failure	A20 门电路故障	安装的某个内存模块可能松动。在升级插槽中重置内存模块。如果问题仍然存在，请从升级插槽中卸下内存模块。
General failure	通用失败	DOS 不能判断错误的原因，一般是因为驱动器中的磁盘未格式化，或格式化成非 DOS 系统。应该重新格式化磁盘。
	通用故障	操作系统无法执行命令。此信息后面通常会有特定的信息，例如，Printer out of paper（打印机纸张用尽）。
Hard disk install failure	硬盘安装失败	这是因为硬盘的电源线或数据线可能未接好或者硬盘跳线设置不当引起的。可以检查一下硬盘的连线是否插好，看看同一根数据线上的两个硬盘的跳线设置是否一样。如果一样，只要将两个硬盘的跳线设置不一样即可(一个设为 Master，另一个设为 Slave)。
Hard disk(s) diagnosis fail	执行硬盘诊断时发生错误	出现这个问题一般就是硬盘内部本身出现硬件故障了，你可以把硬盘放到另一台机子上试一试，如果问题还是没有解决，只能去修一下了。如果硬盘还在包换期内的话，最好还是赶快去换一块!
Hard-disk drive configuration error	硬盘驱动器配置错误	计算机无法识别驱动器的类型。关闭计算机、卸下驱动器并从引导软盘引导计算机，然后再次关闭计算机、重新安装驱动器并重新引导计算机。
Hard-disk drive controller failure 0	硬盘驱动器控制器故障 0	硬盘驱动器无法对计算机发出的命令作出反应。关闭计算机、卸下驱动器并从引导软盘引导计算机，然后再次关闭计算机、重新安装驱动器并重新启动计算机。
Hard-disk drive failure	硬盘驱动器出现故障	硬盘驱动器无法对计算机发出的命令作出反应。关闭计算机、卸下驱动器并从引导软盘引导计算机，然后再次关闭计算机、重新安装驱动器并重新启动计算机即可。
Hard-disk drive read failure	硬盘驱动器读取出现故障	硬盘驱动器可能出现故障。关闭计算机、卸下驱动器并从引导软盘引导计算机，然后再次关闭计算机、重新安装驱动器并重新启动计算机。
Hardware Monitor found an error, enter POWER MANAGEMENT SETUP for details, press F1 to continue, DEL to enter SETUP	监视功能发现错误，进入 POWER MANAGEMENT SETUP 查看详细资料，或按 F1 键继续开机程序，按 DEL 键进入 CMOS 设置	现在好一些的主板都具备硬件的监视功能，用户可以设定主板与 CPU 的温度监视、电压调整器的电压输出准位监视和对各个风扇转速的监视，当上述监视功能在开机时发觉有异常情况，那么便会出现这段话，这时朋友们可以进入 CMOS 设置选择 POWER MANAGEMENT SETUP，在右面的 **Fan Monitor**、**Thermal Monitor** 和 **Voltage Monitor** 查看是哪部分监控发出了异常情况，然后再加以解决。

续表

屏幕提示信息	中文翻译	提示信息解释说明
Incorrect DOS version	DOS 版本不符	输入了一个不同版本的外部命令。用 setver 设置版本或者使用正确的可执行文件。
Insufficient Disk Space	磁盘空间不足	磁盘中已没有可用的空间来拷贝文件或创建文件。可以删除一些无用的文件或更换一个大一点的磁盘。
Insufficient memory	内存不足	没有足够内存来处理用户所输入的命令,一般指基本内存。应删去一些内存驻留的文件或对内存做优化管理。还可以给系统增加更多的内存,以适应应用程序。
Invalid configuration information-please run System Setup program	无效的配置信息-请运行系统设置程序	系统配置信息与硬件配置不匹配,此信息最有可能在安装内存模块之后出现。在系统设置程序中纠正相应的选项。
Invalid directory	非法目录	输入了无效的目录名或不存在的目录名。检查目录的拼法。
Invalid Drive Specification	无效的驱动器定义	根本没有这个驱动器,可能是拼写错误。若是不能指定光驱,可能是没有安装驱动程序。重新安装光驱。
Invalid filename or file not found	无效的文件名或文件未找到	输入的文件名包含了无效字符或通配符,或者将保留设备名用作文件名。利用不同的文件名试试。
Invalid Media, track 0 Bad or Unusable	无效的格式,0 磁道损坏或不可用	一般是磁盘损坏。更换磁盘。
Invalid parameter	无效的参数	在命令行中没有指定正确的参数,或者有重复、禁止的参数。检查命令输入时的拼写或语法。
Invalid partition table	无效的分区表	硬盘分区信息中有错误。应备份所能找到的数据,运行 Fdisk 来重新设置硬盘分区。
Invalid path, not directory, or directory not empty	无效的路径、非目录或目录非空	系统不能定位指定的目录,或者用户输入了文件名来代替目录名,或者目录中包含文件(或子目录),不能被删除。检查目录名的拼法,如果目录为空,那么它可能包含隐含文件,使用 Dir/ah 命令来显示任何可能的隐含文件,用 attrib 改变属性,删除之。
Invalid syntax	无效的语法	系统不能处理用户输入的语法格式。应查阅正确的文件格式再试试。
Keyboard clock line failure	键盘时钟线路出现故障	电缆或连接器可能已松动,或者键盘出现故障。请运行 Dell 诊断程序中的"Keyboard Controller"(键盘控制器)检测程序。
Keyboard controller failure	键盘控制器出现故障	电缆或连接器可能已松动,或者键盘出现故障。重新引导计算机,在引导期间不要触碰键盘或鼠标。
Keyboard data line failure	键盘数据线路出现故障	电缆或连接器可能已松动,或者键盘出现故障。请运行 Dell 诊断程序中的"Keyboard Controller"(键盘控制器)检测程序。
Keyboard error or no keyboard present	键盘错误或者未接键盘	检查一下键盘与主板接口是否接好,如果键盘已经接好,那么就是主板键盘口坏了,主板如尚在保修期内,朋友们可以找经销商或主板厂家进行解决。

续表

屏幕提示信息	中文翻译	提示信息解释说明
Keyboard stuck key failure	键盘上的键被卡住	如果你正在使用外部键盘或小键盘，则可能是电缆或连接器松动，也可能是键盘出现故障。如果你使用的是集成键盘，则可能是键盘出现故障。计算机引导期间，你可能按下了集成键盘或外部键盘上的按键。
Memory address line failure at address, read value expecting value	定址、读取所需的值时，内存地址线路出现故障	安装的某个内存模块可能出现故障或插接不正确。在升级插槽中重置内存模块。如果问题仍然存在，请从升级插槽中卸下内存模块。
Memory allocation error	内存分配错误	你尝试运行的软件与操作系统、另一个应用程序或公用程序发生冲突。关闭计算机并等待30秒，然后重新启动。
Memory data line failure at address, read value expecting value	定址、读取所需的值时，内存数据线路出现故障	安装的某个内存模块可能出现故障或插接不正确。在升级插槽中重置内存模块。如果问题仍然存在，请从升级插槽中卸下内存模块。
Memory double word logic failure at address, read value expecting value	定址、读取所需的值时，内存双字逻辑出现故障	
Memory odd/even logic failure at address, read value expecting value	定址、读取所需的值时，内存奇/偶逻辑出现故障	
Memory test fail	内存检测失败	重新插拔一下内存条，看看是否能解决，出现这种问题一般是因为混插的内存条互相不兼容而引起的，有条件的话去换一条吧！如果使用的只是一根内存条，那么就一定是内存条本身有问题，还是赶快去换一条的好！
Memory write/read failure at address, read value expecting value	定址、读取所需的值时，内存读/写出现故障	
No boot device available	无可用的引导设备	计算机无法找到软盘或硬盘驱动器。如果软盘驱动器是你的引导设备，请确保驱动器中已插入一个可引导软盘。如果将硬盘驱动器用作引导设备，请确保已将其安装、正确插接并分区为引导设备。
No boot sector on hard-disk drive	硬盘驱动器上无引导扇区	操作系统可能已损坏。重新安装操作系统，请参阅随操作系统附带的说明文件。
No fixed disk present	没有硬盘	系统不能检测到硬盘的存在。应检查设置的驱动器类参数，如果不能解决这个问题，那么应该送去修理。
No timer tick interrupt	无计时器嘀嗒信号中断	主机板上的芯片可能出现故障。
Non-System Disk or Disk Error	非系统盘或磁盘错误	系统在当前盘中找不到系统文件。应插入包含系统文件的磁盘，或者重新引导计算机。

※ Appendix B　Screen Prompt Message ※

续表

屏幕提示信息	中文翻译	提示信息解释说明
Non-system disk or disk error	非系统磁盘或磁盘错误	A 驱动器中的软盘或你的硬盘驱动器中未安装可引导操作系统的文件。如果你尝试从软盘引导，请更换另一张具有可引导操作系统的软盘。
Not a boot diskette	不是引导软盘	软盘上无操作系统。使用一张包含操作系统的软盘引导系统。
Not enough memory	内存不足	
NOT READY, READING DRIVE X	驱动器 X 未准备好	在指定的驱动器中没有盘或门没关。插入磁盘到指定驱动器或关上驱动器门。
Optional ROM bad checksum	可选 ROM 校验和错误	可选 ROM 出现明显错误。
Override enable-defaults loaded	当前 CMOS 设定无法启动系统，载入 BIOS 中的预设值以便启动系统	一般是 CMOS 内的设定出现错误，朋友们只要进入 CMOS 设置后选择 LOAD SETUP DEFAULTS 载入系统原来的设定值然后重新启动即可解决这一问题。
Press ESC to skip memory test	正在进行内存检查，可按 ESC 键跳过	这是因为在 CMOS 内没有设定跳过存储器的第二、三、四次测试，开机就会执行四次内存测试，当然你也可以按 ESC 键结束内存检查，不过每次开机后都要这样做实在太麻烦了，你可以进入 CMOS 设置后选择 BIOS FEATURES SETUP，将其中的 Quick Power On Self Test 设为 Enabled 开启，存储后重新启动即可解决。
Program too big to fit in memory	程序太大不能载入内存	参见 Insufficient memory。
Required Parameter missing	缺少必要的参数	见 Invalid Parameter。
Resuming from disk, press TAB to show POST screen	从硬盘恢复开机，按 TAB 显示开机自检画面	这是因为市场中有的主板在 BIOS 设置中提供了 Suspend to disk（将硬盘挂起）功能，如果我们使用了 Suspend to disk 的方式来关机，那么我们在下次开机时，屏幕上就会显示此提示消息的。
Secondary slave hard fail	检测从盘失败	可能是 CMOS 设置不当引起的。比如说 IDE 设备中并没有连接从盘，但在 CMOS 里设置有从盘，那么就会出现这个错误。如出现这个错误，朋友们可以进入 CMOS 设置，选择 IDE HDD AUTO DETECTION 进行硬盘自动侦测。如果使用上述方法问题没有解决，那就可能是硬盘的电源线、数据线未接好或者硬盘跳线设置不当。
Sector not found	未找到扇区	操作系统无法找到软盘或硬盘驱动器上的某个扇区。你的软盘或硬盘驱动器上可能有坏扇区或损坏的 FAT。运行相应的公用程序，检查软盘或硬盘驱动器上的文件结构。如果大量的扇区出现故障，请备份数据（如果有可能），然后重新格式化软盘或硬盘驱动器。

续表

屏幕提示信息	中文翻译	提示信息解释说明
Seek error	寻道错误	操作系统无法找到软盘或硬盘驱动器上的特定磁道。如果错误与软盘驱动器有关，请尝试使用另一张软盘。
Shutdown failure	关闭系统失败	主机板上的芯片可能出现故障。运行 Dell 诊断程序中的"System Set"（系统设置）检测程序。
Time-of-day clock lost power	计时时钟电源中断	存储在 NVRAM 中的数据已损坏。将计算机连接至电源插座，为电池充电。如果充电后问题仍然存在，请尝试恢复数据。要恢复数据，请按 <Fn><F1> 组合键进入系统设置程序，然后立即退出。如果此信息再次出现，请致电 Dell 寻求技术帮助。
Time-of-day clock stopped	计时时钟停止	维持 NVRAM 中数据的备用电池的电能可能已用尽。将计算机连接至电源插座，为电池充电。
Time-of-day not set - please run the System Setup program	未设置时间 – 请运行系统设置程序	存储在系统设置程序中的时间或日期与系统时钟不符。纠正"Date"（日期）与"Time"（时间）选项的设置。
Timer chip counter 2 failed	计时器芯片计数器 2 出现故障	主机板上的芯片可能出现故障。运行 Dell 诊断程序中的"System Set"（系统设置）检测程序。
Too many open files	打开的文件太多	超过系统规定的打开文件数目。应在 Config.sys 文件中用 Files 命令增加最大数目，并重新引导计算机。
Unexpected interrupt in protected mode	保护模式中出现意外中断	键盘控制器可能出现故障，或者安装的某个内存模块松动。运行 Dell 诊断程序中的"System Memory"（系统内存）检测程序和"Keyboard Controller"（键盘控制器）检测程序。
Unrecognized command in CONFIG.SYS	config.sys 中有不可辨认的命令	在引导系统时，不能识别 Config.sys 文件中的命令，应编辑 config.sys 文件，修正无效的行。
Warning: Battery is critically low	警告：电池电量过低	电池中的电量即将耗尽。更换电池，或者将计算机连接至电源插座。否则，请激活状态保持模式或关闭计算机。
Write fault error	写失败错误	系统不能在磁盘上写数据。将磁盘取下再重新插好试试，仍然不行则运行 Chkdsk 或 Scandisk，如磁盘不能恢复，扔掉它。
Write protect error	写保护错误	磁盘上有写保护，取下磁盘，去掉写保护即可。

Appendix C Index of Basic Vocabulary

abacus /ˈæbəkəs/ n.　算盘
all-in-one　一体机
analog /ˈænəlɒg/ adj.　模拟的
analytical engine　分析机
animation /ˌænɪˈmeɪʃn/ n.　动画制作
anti-aliasing　抗锯齿的
antivirus program　防病毒程序
appropriate /əˈprəupriət/ adj.　恰当的；合适的
arcade machine　街机
architecture /ˈɑːkɪtektʃə/ n.　体系结构
arithmetic /əˈrɪθmətɪk/ n.　算术
arithmetic logic unit　算术逻辑单元
artificial language　人工语言
aspect /ˈæspekt/ n.　方面；外观
barrier /ˈbæriə/ n.　障碍物
（be）along with　跟随；伴随
（be）not foolproof　不是万无一失
（be）dedicated to　用于……
brief /briːf/ adj.　简要的
Blu-ray disc　蓝光光盘
Border Gateway Protocol　边界网关协议
Caesar cipher　恺撒密码
capacity /kəˈpæsəti/ n.　容量
category /ˈkætəgəri/ n.　类（别）
cathode ray tube　阴极射线管
central processing unit　中央处理器
characteristic /ˌkærəktəˈrɪstɪk/ n.　特点；特征
class-based　基于类的
client/server architecture　客户端/服务器体系结构
cocoa touch　苹果公司开发的可触摸界面开发框架

cognitive /ˈkɒgnətɪv/ *adj.* 认知的
combine /kəmˈbaɪn/ *v.* 结合
command /kəˈmɑːnd/ *n.* 命令；指令
commerce /ˈkɒmɜːs/ *n.* 贸易
component /kəmˈpəunənt/ *n.* 组件；元件；部分
computer network 计算机网络
computer-assisted learning 计算机辅助学习
confidential /kɒnfɪˈdenʃl/ *adj.* 机密的
confidentiality /ˌkɒnfɪˌdenʃɪˈæləti/ *n.* 机密
considerable /kənˈsɪdərəbl/ *adj.* 相当多的；重要的
consist of 由……组成
console game 主机游戏
convey /kənˈveɪ/ *v.* 传送；输送
critical issue 关键问题
data encryption 数据加密
database management application 数据库管理应用
declarative form 声明形式
default route 默认路由
demonstration /ˌdemənsˈtreɪʃn/ *n.* 示范
desktop publishing 桌面出版
digital /ˈdɪdʒɪtl/ *adj.* 数字的；数码的
digital forensics 数字取证
dramatically /drəˈmætɪkli/ *adv.* 戏剧地；突然；巨大地
duplication /ˌdjuːplɪˈkeɪʃn/ *n.* （不必要的）重复；复制
electrical engineering 电气工程
electromechanical /ɪˌlektrəumɪˈkænɪkl/ *adj.* 机电的
electronic data processing 电子数据处理
emerge /iˈmɜːdʒ/ *v.* 浮现；摆脱；暴露
encryption /ɪnˈkrɪpʃn/ *n.* 加密
enhance /ɪnˈhɑːns/ *v.* 提高；增强
ensure /ɪnˈʃuə/ *v.* 保证；确保
entity–relationship model 实体关系模型
essential /ɪˈsenʃl/ *n.* 要素；实质
e-tail /ˈiːteɪ/ *n.* 电子零售

evidence /ˈevɪdəns/ n. 证据
evolve from... to... 演进
exceed /ɪkˈsiːd/ v. 超过；超越
explosive /ɪkˈspləʊsɪv/ adj. 突增的；猛增的
exponential /ˌekspəˈnenʃl/ adj. 指数的
facilitate /fəˈsɪlɪteɪt/ v. 使便利；减轻……的困难
fiber optical cable 光纤
fidelity /fɪˈdeləti/ n. 保真；逼真度
financial /faɪˈnænʃl/ adj. 金融的；财务的
firewall /ˈfaɪəwɔːl/ n. 防火墙
flat file 扁平文件
foreseeable /fɔːˈsiːəbl/ adj. 可以预知的；能预见的
game engine 游戏引擎
geographical /ˌdʒiːəˈɡræfɪkl/ adj. 地理的
go through 通过；复习
graphics /ˈɡræfɪks/ n. 图形；绘图
guaranteed /ˌɡærənˈtiːd/ adj. 保证的；规定的
hacking program 黑客程序
handheld /ˈhændheld/ adj. 掌上型的
haptic /ˈhæptɪk/ adj. 触觉的
haptic technology 触感技术
hardware accelerated 硬件加速
hardware-independent 与硬件无关的
hierarchical /ˌhaɪəˈrɑːkɪkl/ adj. 按等级划分的
high-level 高级
high-volume 大容量
hypertext /ˈhaɪpətekst/ n. 超文本
illustration /ˌɪləˈstreɪʃn/ n. 插图；图解
imperative programming language 命令式编程语言
incremental /ˌɪŋkrəˈmentl/ adj. 增加的；增值的
individual user 个人用户
inexpensive /ˌɪnɪkˈspensɪv/ adj. 物美价廉的
inexpensively /ˌɪnɪkˈspensɪvli/ adv. 廉价地
input /ˈɪnpʊt/ n. 输入

input device 输入设备
instruction /in'strʌkʃn/ n. 指令；命令；操作指南
insulate M from N 把 M 和 N 分隔开
integrity /in'tegrəti/ n. 完整性
internal /ɪn'tɜ:nl/ adj. 内部的
Internet protocol suite 互联网协议族
isolation /ˌaɪsə'leɪʃn/ n. 隔离；孤立状态
landmark /'lændmɑ:k/ n. 里程碑
leakage /'li:kɪdʒ/ n. 泄漏
leap /li:p/ n. 跳跃；剧增
leased line 专线
liquid crystal display 液晶显示屏
machine code 机器码
magnetic tape 磁带
manipulate /mə'nɪpjuleɪt/ v. 操作；使用
manufacturer /ˌmænju'fæktʃərə/ n. 制造商
marketplace /'mɑ:kɪtpleɪs/ n. 市场
mechanical calculator 机械计算器
mechanism /'mekənɪzəm/ n. 机制；构造
memory /'meməri/ n. 内存
microprocessor technology 微处理器技术
middleware /'mɪdlweə/ n. 中间件
modifiable /'mɒdɪfaɪəbl/ adj. 可修饰的
motherboard /'mʌðəbɔ:d/ n. 主板
multi-paradigm language 多范式语言
multiplication /ˌmʌltɪplɪ'keɪʃn/ n. 乘法
nanometer /'nænəumi:tə/ n. 纳米
nanotube /'nænəutju:b/ n. 纳米碳管
neuron /'njʊərɒn/ n. 神经元
object-oriented programming 面向对象编程
object-relational system 对象关系系统
Official Secrets Act 《官方保密法》
Open Handset Alliance 开放手机联盟
output /'aʊtpʊt/ n. 输出

output device　输出设备
paradigm /'pærədaɪm/ n.　典范；样式
particle /'pɑːtɪkl/ n.　颗粒
peer-to-peer　对等网
peripheral /pə'rɪfərəl/ adj.　外围的；与计算机相连的
phenomenon /fə'nɒmɪnən/ n.　现象
photograph /'fəʊtəgrɑːf/ n.　照片
pipeline /'paɪplaɪn/ n.　管道
player piano　自动钢琴
plummeting /'plʌmɪtɪŋ/ adj.　直线下降的
potential /pə'tenʃl/ adj.　潜在的
predict /prɪ'dɪkt/ v.　预言；预报
pre-emptive scheduling　占先调度
privacy /'prɪvəsi/ n.　隐私；不受公众干扰的状态
program specification　程序规范；程序规格说明
programming language　编程语言
proxy server　代理服务器
quantum /'kwɒntəm/ n.　量子
random /'rændəm/ adj.　随机的；随意的　n. 随意
range from... to...　范围从……到……
raw /rɔː/ adj.　（材料，物质）自然状态的；未加工的；未经处理的
recipient /rɪ'sɪpiənt/ n.　接受者
reinforce /ˌriːɪn'fɔːs/ v.　加固；给……加强力量（或装备）
revolutionize /ˌrevə'luːʃənaɪz/ v.　彻底改变
roughly /'rʌfli/ adv.　大约；大致
routing table　路由表
schedule /'ʃedjuːl/ v.　调度
schema /'skiːmə/ n.　提要；纲要
sci-fi　科幻
scripting language　脚本语言
security testing　安全性测试
sequential /sɪ'kwenʃl/ adj.　按次序的；顺序的；序列的
server-side　服务器端
servo /'sɜːvəʊ/ n.　伺服系统

simulation /ˌsɪmjuˈleɪʃn/ n. 仿真；模拟
simultaneously /ˌsɪmlˈteɪniəsli/ adj. 同时地
single-user system 单用户系统
spreadsheet /ˈspredʃiːt/ n. 电子表格
spyware /ˈspaɪweə/ n. 间谍软件；间谍程序
storage device 存储设备
streaming multimedia 流媒体
structured document-oriented 面向结构化文档的
stylus /ˈstaɪləs/ n. 触笔；铁笔
substantially /səbˈstænʃəli/ adv. 基本上；大体上
suite /swiːt/ n. 套装软件
supersede /ˌsuːpəˈsiːd/ v. 取代
symbolic /sɪmˈbɒlɪk/ adj. 符号的；象征的
syntax and semantics 语法和语义
tablet computer 平板电脑
telecommunications /ˌtelikəˌmjuːnɪkeɪʃnz/ n. 电信
telepathy /təˈlepəθi/ n. 心灵感应
terabyte /ˈterəbaɪt/ n. 万亿字节；太字节
third-person shooter game 第三人称射击游戏
token-ring network 令牌环网
topology /təˈpɒlədʒi/ n. 拓扑结构
touchscreen mobile device 触摸屏移动设备
trajectory /trəˈdʒekətəri/ n. 轨迹
transaction processing 事务处理
transistor /trænˈzɪstə/ n. 晶体管
tunnel /ˈtʌnl/ v. 挖隧道
unauthorized access 非授权存取
upgrade /ˈʌpɡreɪd/ n. 升级；上升
vacuum tube 真空管
view /vjuː/ n. 视图
virus /ˈvaɪrəs/ n. 病毒
word processing 文字处理
word processor 文字处理软件
worm /wɜːm/ n. 蠕虫

Appendix D Concise Explanation of Common Abbreviations

4GL	编程发展第四代语言（Fourth-Generation Language）
ADHD	多动症
ALGOL	国际代数语言，是计算机发展史上首批产生的高级语言
ALU	算术逻辑单元
Android	Google 公司推出的基于 Linux 的移动设备操作系统
ANSI	美国国家标准化组织
API	应用程序编程接口
APL	Array Processing Language 的缩写，阵列处理语言
B2B	企业对企业的电子商务模式
B2C	企业对消费者的电子商务模式
B2E	企业对雇员的电子商务模式，雇员是一种资源
B2G	企业对政府的电子商务模式
BGP	边界网关协议
C2B	消费者对企业的电子商务模式
C2C	消费者对消费者的电子商务模式
CD	光碟
CERN	欧洲核子研究中心
COBOL	面向商业的通用语言
CODASYL DBTG	数据系统语言会议数据库任务组
CP/M	控制程序微机
CPU	中央处理器
DB/DC	数据库与数据通信
DB2	IBM 关系型数据库管理系统
DBMS	数据库管理系统
DCL	数据控制语言
DDL	数据定义语言
DEC	一家老牌电脑公司，后被康柏电脑收购
DHCP	动态主机配置协议
DirectX	微软公司创建的编程的接口
DML	数据操作语言
DS	便携式游戏机，即任天堂的 NDS
DVD	数字视频光盘
Dwarf Fortress	矮人要塞，一款游戏
EC	电子贸易

ENIAC	电子数字积分计算机
ERP	企业资源计划
FDD	软盘驱动器
FORTRAN	Formula Translation 的缩写，是一种编程语言
FTP	文件传输协议
G2B	政府对企业的电子商务模式
G2C	政府与公众的电子商务模式，也就是电子政务
GPS	全球定位系统
GUI	图形用户界面
Havok	一种游戏引擎，游戏开发中使用
HDD	硬盘驱动器
HTML	超文本标记语言
HTTP	超文本传输协议
IBM	美国国际商用机器公司
ICT	通信和信息技术
IDS	信息显示系统，入侵检测系统
IMS	信息管理系统（[美]IBM 公司软件）
Intel	美国英特尔公司，以生产 CPU 芯片著称
iOS	苹果公司为其移动设备开发的操作系统
IoT	物联网
iPad	苹果公司开发的平板电脑
iPhone	苹果公司开发的智能手机
iPod Touch	苹果公司推出的一种大容量 MP3 音乐播放器
IPv4	互联网通信协议第 4 版
IPv6	互联网通信协议第 6 版
ISO	国际标准化组织
ISP	互联网服务提供商
LAN	局域网
LCD	液晶显示器
Longhorn	Windows 操作系统的下一个版本 长角
MAN	城域网
MCA	微通道结构总线技术
Micro-SD	一种迷你型的 SD 卡，当前市场上的手机内存卡就是 micro-SD 卡
Mi-Fi	一个便携式宽带无线装置
MIT	麻省理工学院
MRP	物资需求计划，材料需求计划
NASA	美国航空航天局

NDA	保密协议
NPL	不良债权，是指银行放款债权
ODD	光盘驱动器
OLAP	联机分析技术（On-Line Analytical Processing）
OODBMS	面向对象的数据库管理系统
OpenGL	一个专业的 3D 程序接口
Oracle	甲骨文公司（Oracle）是世界上最大的企业软件公司之一
OSX	苹果公司开发的苹果电脑操作系统
OS/2	由 IBM 单独开发的一套操作系统
OSI	开放式系统互联参考模型
PC	个人电脑
PDA	个人数字助理（Personal Digital Assistant）
petaFLOPS	每秒千万亿次浮点运算
QR code	二维码
RAM	随机存储器
RFID	无线射频识别
SD	安全数字卡，很多人将其叫作内存卡，多用于手机，相机中
SEM	搜索引擎营销
SIMD	单指令流多数据流
Spore	孢子，一款游戏
SQL	结构化查询语言
SSE	数据流单指令多数据扩展指令集，系统监视设备
TB	数据容量单位，等于 1 024 GB
TCP/IP	传输控制协议/因特网互联协议
Total War	全面战争，一款游戏
UCSD	圣地亚哥加州大学
UK	联合王国
UPS	不间断电源
USB	通用串行总线
VDU	视频显示单元
Vista	微软继 Windows XP 后推出的新视窗操作系统
VPN	虚拟专用网
WAN	广域网
Wi-Fi	无线上网技术
WinFS	Windows 未来存储
WWW	万维网
XML	可扩展标记语言